MAKING SENSE OF
DATA I

MAKING SENSE OF DATA I
A Practical Guide to Exploratory Data Analysis and Data Mining

Second Edition

GLENN J. MYATT
WAYNE P. JOHNSON

Published by John Wiley & Sons, Inc., Hoboken, New Jersey
Published simultaneously in Canada

For general information on our other products and services or for technical support, please contact
our Customer Care Department within the United States at (800) 762-2974, outside the United States
at (317) 572-3993 or fax (317) 572-4002.

Wiley also publishes its books in a variety of electronic formats. Some content that appears in print
may not be available in electronic formats. For more information about Wiley products, visit our web
site at www.wiley.com.

Library of Congress Cataloging-in-Publication Data:

Myatt, Glenn J., 1969–
    [Making sense of data]
    Making sense of data I : a practical guide to exploratory data analysis and data mining /
Glenn J. Myatt, Wayne P. Johnson. – Second edition.
        pages cm
    Revised edition of: Making sense of data. c2007.
    Includes bibliographical references and index.
    ISBN 978-1-118-40741-7 (paper)
    1. Data mining.   2. Mathematical statistics.   I. Johnson, Wayne P.   II. Title.
    QA276.M92 2014
    006.3′12–dc23

                                                                    2014007303

ISBN: 978-1-118-40741-7

10 9 8 7 6 5 4 3 2 1

CONTENTS

PREFACE

An unprecedented amount of data is being generated at increasingly rapid rates in many disciplines. Every day retail companies collect data on sales transactions, organizations log mouse clicks made on their websites, and biologists generate millions of pieces of information related to genes. It is practically impossible to make sense of data sets containing more than a handful of data points without the help of computer programs. Many free and commercial software programs exist to sift through data, such as spreadsheet applications, data visualization software, statistical packages and scripting languages, and data mining tools. Deciding what software to use is just one of the many questions that must be considered in exploratory data analysis or data mining projects. Translating the raw data collected in various ways into actionable information requires an understanding of exploratory data analysis and data mining methods and often an appreciation of the subject matter, business processes, software deployment, project management methods, change management issues, and so on.

The purpose of this book is to describe a practical approach for making sense out of data. A step-by-step process is introduced, which is designed to walk you through the steps and issues that you will face in data analysis or data mining projects. It covers the more common tasks relating to the analysis of data including (1) how to prepare data prior to analysis, (2) how to generate summaries of the data, (3) how to identify non-trivial facts, patterns, and relationships in the data, and (4) how to create models from the data to better understand the data and make predictions.

The process outlined in the book starts by understanding the problem you are trying to solve, what data will be used and how, who will use the information generated, and how it will be delivered to them, and the specific and measurable success criteria against which the project will be evaluated.

The type of data collected and the quality of this data will directly impact the usefulness of the results. Ideally, the data will have been carefully collected to answer the specific questions defined at the start of the project. In practice, you are often dealing with data generated for an entirely different purpose. In this situation, it is necessary to thoroughly understand and prepare the data for the new questions being posed. This is often one of the most time-consuming parts of the data mining process where many issues need to be carefully adressed.

The analysis can begin once the data has been collected and prepared. The choice of methods used to analyze the data depends on many factors, including the problem definition and the type of the data that has been collected. Although many methods might solve your problem, you may not know which one works best until you have experimented with the alternatives. Throughout the technical sections, issues relating to when you would apply the different methods along with how you could optimize the results are discussed.

After the data is analyzed, it needs to be delivered to your target audience. This might be as simple as issuing a report or as complex as implementing and deploying new software to automatically reapply the analysis as new data becomes available. Beyond the technical challenges, if the solution changes the way its intended audience operates on a daily basis, it will need to be managed. It will be important to understand how well the solution implemented in the field actually solves the original business problem.

Larger projects are increasingly implemented by interdisciplinary teams involving subject matter experts, business analysts, statisticians or data mining experts, IT professionals, and project managers. This book is aimed at the entire interdisciplinary team and addresses issues and technical solutions relating to data analysis or data mining projects. The book also serves as an introductory textbook for students of any discipline, both undergraduate and graduate, who wish to understand exploratory data analysis and data mining processes and methods.

The book covers a series of topics relating to the process of making sense of data, including the data mining process and how to describe data table elements (i.e., observations and variables), preparing data prior to analysis, visualizing and describing relationships between variables, identifying and making statements about groups of observations, extracting interesting rules, and building mathematical models that can be used to understand the data and make predictions.

The book focuses on practical approaches and covers information on how the techniques operate as well as suggestions for when and how to use the different methods. Each chapter includes a "Further Reading" section that highlights additional books and online resources that provide background as well as more in-depth coverage of the material. At the end of selected chapters are a set of exercises designed to help in understanding the chapter's material. The appendix covers a series of practical tutorials that make use of the freely available Traceis software developed to accompany the book, which is available from the book's website: http://www.makingsenseofdata.com; however, the tutorials could be used with other available software. Finally, a deck of slides has been developed to accompany the book's material and is available on request from the book's authors.

The authors wish to thank Chelsey Hill-Esler, Dr. McCullough, and Vinod Chandnani for their help with the book.

CHAPTER 1

INTRODUCTION

1.1 OVERVIEW

Almost every discipline from biology and economics to engineering and marketing measures, gathers, and stores data in some digital form. Retail companies store information on sales transactions, insurance companies keep track of insurance claims, and meteorological organizations measure and collect data concerning weather conditions. Timely and well-founded decisions need to be made using the information collected. These decisions will be used to maximize sales, improve research and development projects, and trim costs. Retail companies must determine which products in their stores are under- or over-performing as well as understand the preferences of their customers; insurance companies need to identify activities associated with fraudulent claims; and meteorological organizations attempt to predict future weather conditions.

Data are being produced at faster rates due to the explosion of internet-related information and the increased use of operational systems to collect business, engineering and scientific data, and measurements from sensors or monitors. It is a trend that will continue into the foreseeable future. The challenges of handling and making sense of this information are significant

Making Sense of Data I: A Practical Guide to Exploratory Data Analysis and Data Mining,
Second Edition. Glenn J. Myatt and Wayne P. Johnson.
© 2014 John Wiley & Sons, Inc. Published 2014 by John Wiley & Sons, Inc.

because of the increasing volume of data, the complexity that arises from the diverse types of information that are collected, and the reliability of the data collected.

The process of taking raw data and converting it into meaningful information necessary to make decisions is the focus of this book. The following sections in this chapter outline the major steps in a data analysis or data mining project from defining the problem to the deployment of the results. The process provides a framework for executing projects related to data mining or data analysis. It includes a discussion of the steps and challenges of (1) defining the project, (2) preparing data for analysis, (3) selecting data analysis or data mining approaches that may include performing an optimization of the analysis to refine the results, and (4) deploying and measuring the results to ensure that any expected benefits are realized. The chapter also includes an outline of topics covered in this book and the supporting resources that can be used alongside the book's content.

1.2 SOURCES OF DATA

There are many different sources of data as well as methods used to collect the data. Surveys or polls are valuable approaches for gathering data to answer specific questions. An interview using a set of predefined questions is often conducted over the phone, in person, or over the internet. It is used to elicit information on people's opinions, preferences, and behavior. For example, a poll may be used to understand how a population of eligible voters will cast their vote in an upcoming election. The specific questions along with the target population should be clearly defined prior to the interviews. Any bias in the survey should be eliminated by selecting a random sample of the target population. For example, bias can be introduced in situations where only those responding to the questionnaire are included in the survey, since this group may not be representative of a random sample of the entire population. The questionnaire should not contain leading questions—questions that favor a particular response. Other factors which might result in segments of the total population being excluded should also be considered, such as the time of day the survey or poll was conducted. A well-designed survey or poll can provide an accurate and cost-effective approach to understanding opinions or needs across a large group of individuals without the need to survey everyone in the target population.

Experiments measure and collect data to answer specific questions in a highly controlled manner. The data collected should be reliably measured; in other words, repeating the measurement should not result in substantially different values. Experiments attempt to understand cause-and-effect phenomena by controlling other factors that may be important. For example, when studying the effects of a new drug, a double-blind study is typically used. The sample of patients selected to take part in the study is divided into two groups. The new drug is delivered to one group, whereas a placebo (a sugar pill) is given to the other group. To avoid a bias in the study on the part of the patient or the doctor, neither the patient nor the doctor administering the treatment knows which group a patient belongs to. In certain situations it is impossible to conduct a controlled experiment on either logistical or ethical grounds. In these situations a large number of observations are measured and care is taken when interpreting the results. For example, it would not be ethical to set up a controlled experiment to test whether smoking causes health problems.

As part of the daily operations of an organization, data is collected for a variety of reasons. *Operational databases* contain ongoing business transactions and are accessed and updated regularly. Examples include supply chain and logistics management systems, customer relationship management databases (CRM), and enterprise resource planning databases (ERP). An organization may also be automatically monitoring operational processes with sensors, such as the performance of various nodes in a communications network. A *data warehouse* is a copy of data gathered from other sources within an organization that is appropriately prepared for making decisions. It is not updated as frequently as operational databases. Databases are also used to house historical polls, surveys, and experiments. In many cases data from in-house sources may not be sufficient to answer the questions now being asked of it. In these cases, the internal data can be augmented with data from other sources such as information collected from the web or literature.

1.3 PROCESS FOR MAKING SENSE OF DATA

1.3.1 Overview

Following a predefined process will ensure that issues are addressed and appropriate steps are taken. For exploratory data analysis and data mining projects, you should carefully think through the following steps, which are summarized here and expanded in the following sections:

1. **Problem definition and planning**: The problem to be solved and the projected deliverables should be clearly defined and planned, and an appropriate team should be assembled to perform the analysis.

FIGURE 1.1 Summary of a general framework for a data analysis project.

 2. **Data preparation**: Prior to starting a data analysis or data mining project, the data should be collected, characterized, cleaned, transformed, and partitioned into an appropriate form for further processing.
 3. **Analysis**: Based on the information from steps 1 and 2, appropriate data analysis and data mining techniques should be selected. These methods often need to be optimized to obtain the best results.
 4. **Deployment**: The results from step 3 should be communicated and/or deployed to obtain the projected benefits identified at the start of the project.

Figure 1.1 summarizes this process. Although it is usual to follow the order described, there will be interactions between the different steps that may require work completed in earlier phases to be revised. For example, it may be necessary to return to the data preparation (step 2) while implementing the data analysis (step 3) in order to make modifications based on what is being learned.

1.3.2 Problem Definition and Planning

The first step in a data analysis or data mining project is to describe the problem being addressed and generate a plan. The following section addresses a number of issues to consider in this first phase. These issues are summarized in Figure 1.2.

Problem definition and planning
- Identify the problem or need to be addressed
- List the project's deliverables
- Generate success factors
- Understand each resource and other limitations
- Put together an appropriate team
- Create a plan
- Perform a costs/benefits analysis

FIGURE 1.2 Summary of some of the issues to consider when defining and planning a data analysis project.

It is important to document the business or scientific problem to be solved along with relevant background information. In certain situations, however, it may not be possible or even desirable to know precisely the sort of information that will be generated from the project. These more open-ended projects will often generate questions by exploring large databases. But even in these cases, identifying the business or scientific problem driving the analysis will help to constrain and focus the work. To illustrate, an e-commerce company wishes to embark on a project to redesign their website in order to generate additional revenue. Before starting this potentially costly project, the organization decides to perform data analysis or data mining of available web-related information. The results of this analysis will then be used to influence and prioritize this redesign. A general problem statement, such as "make recommendations to improve sales on the website," along with relevant background information should be documented.

This broad statement of the problem is useful as a headline; however, this description should be divided into a series of clearly defined deliverables that ultimately solve the broader issue. These include: (1) categorize website users based on demographic information; (2) categorize users of the website based on browsing patterns; and (3) determine if there are any relationships between these demographic and/or browsing patterns and purchasing habits. This information can then be used to tailor the site to specific groups of users or improve how their customers purchase based on the usage patterns found in the analysis. In addition to understanding what type of information will be generated, it is also useful to know how it will be delivered. Will the solution be a report, a computer program to be used for making predictions, or a set of business rules? Defining these deliverables will set the expectations for those working on the project and for its stakeholders, such as the management sponsoring the project.

The success criteria related to the project's objective should ideally be defined in ways that can be measured. For example, a criterion might be to increase revenue or reduce costs by a specific amount. This type of criteria can often be directly related to the performance level of a computational model generated from the data. For example, when developing a computational model that will be used to make numeric projections, it is useful to understand the required level of accuracy. Understanding this will help prioritize the types of methods adopted or the time or approach used in optimizations. For example, a credit card company that is losing customers to other companies may set a business objective to reduce the turnover rate by 10%. They know that if they are able to identify customers likely to switch to a competitor, they have an opportunity to improve retention through additional marketing. To identify these customers, the company decides to build a predictive model and the accuracy of its predictions will affect the level of retention that can be achieved.

It is also important to understand the consequences of answering questions incorrectly. For example, when predicting tornadoes, there are two possible prediction errors: (1) incorrectly predicting a tornado would strike and (2) incorrectly predicting there would be no tornado. The consequence of scenario (2) is that a tornado hits with no warning. In this case, affected neighborhoods and emergency crews would not be prepared and the consequences might be catastrophic. The consequence of scenario (1) is less severe than scenario (2) since loss of life is more costly than the inconvenience to neighborhoods and emergency services that prepared for a tornado that did not hit. There are often different business consequences related to different types of prediction errors, such as incorrectly predicting a positive outcome or incorrectly predicting a negative one.

There may be restrictions concerning what resources are available for use in the project or other constraints that influence how the project proceeds, such as limitations on available data as well as computational hardware or software that can be used. Issues related to use of the data, such as privacy or legal issues, should be identified and documented. For example, a data set containing personal information on customers' shopping habits could be used in a data mining project. However, if the results could be traced to specific individuals, the resulting findings should be anonymized. There may also be limitations on the amount of time available to a computational algorithm to make a prediction. To illustrate, suppose a web-based data mining application or service that dynamically suggests alternative products to customers while they are browsing items in an online store is to be developed. Because certain data mining or modeling methods take a long time to generate an answer, these approaches should be avoided if suggestions must be generated rapidly (within a few seconds) otherwise the customer will become frustrated and shop elsewhere. Finally, other restrictions relating to business issues include the *window of opportunity* available for the deliverables. For example, a company may wish to develop and use a predictive model to prioritize a new type of shampoo for testing. In this scenario, the project is being driven by competitive intelligence indicating that another company is developing a similar shampoo and the company that is first to market the product will have a significant advantage. Therefore, the time to generate the model may be an important factor since there is only a small window of opportunity based on business considerations.

Cross-disciplinary teams solve complex problems by looking at the data from different perspectives. Because of the range of expertise often required, teams are essential—especially for large-scale projects—and it is helpful to consider the different roles needed for an interdisciplinary team. A *project leader* plans and directs a project, and monitors its results. *Domain experts* provide specific knowledge of the subject matter or business problems, including (1) how the data was collected, (2) what the data values mean, (3) the accuracy of the data, (4) how to interpret the results of the analysis, and (5) the business issues being addressed by the project. *Data analysis/mining experts* are familiar with statistics, data analysis methods, and data mining approaches as well as issues relating to data preparation. An *IT specialist* has expertise in integrating data sets (e.g., accessing databases, joining tables, pivoting tables) as well as knowledge of software and hardware issues important for implementation and deployment. *End users* use information derived from the data routinely or from a one-off analysis to help them make decisions. A single member of the team may take on multiple roles such as the role of project leader and data analysis/mining expert, or several individuals may be responsible for a single role. For example, a team may include multiple subject matter experts, where one individual has knowledge of how the data was measured and another has knowledge of how it can be interpreted. Other individuals, such as the project sponsor, who have an interest in the project should be included as interested parties at appropriate times throughout the project. For example, representatives from the finance group may be involved if the solution proposes a change to a business process with important financial implications.

Different individuals will play active roles at different times. It is desirable to involve all parties in the project definition phase. In the data preparation phase, the IT expert plays an important role in integrating the data in a form that can be processed. During this phase, the data analysis/mining expert and the subject matter expert/business analyst will also be working closely together to clean and categorize the data. The data analysis/mining expert should be primarily responsible for ensuring that the data is transformed into a form appropriate for analysis. The analysis phase is primarily the responsibility of the data analysis/mining expert with input from the subject matter expert or business analyst. The IT expert can provide a valuable hardware and software support role throughout the project and will play a critical role in situations where the output of the analysis is to be integrated within an operational system.

With cross-disciplinary teams, communicating within the group may be challenging from time-to-time due to the disparate backgrounds of the members of the group. A useful way of facilitating communication is to define and share glossaries defining terms familiar to the subject matter experts or to the data analysis/data mining experts. Team meetings to share information are also essential for communication purposes.

The extent of the project plan depends on the size and scope of the project. A timetable of events should be put together that includes the preparation, implementation, and deployment phases (summarized in Sections 1.3.3, 1.3.4, and 1.3.5). Time should be built into the timetable for reviews after each phase. At the end of the project, a valuable exercise that provides insight for future projects is to spend time evaluating what did and did not work. Progress will be iterative and not strictly sequential, moving between phases of the process as new questions arise. If there are high-risk steps in the process, these should be identified and contingencies for them added to the plan. Tasks with dependencies and contingencies should be documented using timelines or standard project management support tools such as Gantt charts. Based on the plan, budgets and success criteria can be developed to compare costs against benefits. This will help determine the feasibility of the project and whether the project should move forward.

1.3.3 Data Preparation

In many projects, understanding the data and getting it ready for analysis is the most time-consuming step in the process, since the data is usually integrated from many sources, with different representations and formats. Figure 1.3 illustrates some of the steps required for preparing a data set. In situations where the data has been collected for a different purpose, the data will need to be transformed into an appropriate form for analysis. For example, the data may be in the form of a series of documents that requires it to be extracted from the text of the document and converted to a tabular form that is amenable for data analysis. The data should be prepared to mirror as closely as possible the target population about which new questions will be asked. Since multiple sources of data may be used, care must be taken not to introduce errors when these sources are brought together. Retaining information about the source is useful both for bookkeeping and for interpreting the results.

Data Preparation

- Access and combine data tables
- Summarize data
- Look for errors
- Transform data
- Segment data

FIGURE 1.3   Summary of steps to consider when preparing the data.

It is important to characterize the types of attributes that have been collected over the different items in the data set. For example, do the attributes represent discrete categories such as color or gender or are they numeric values of attributes such as temperature or weight? This categorization helps identify unexpected values. In looking at the numeric attribute *weight* collected for a set of people, if an item has the value "low" then we need to either replace this erroneous value or remove the entire record for that person. Another possible error occurs in values for observations that lie outside the typical range for an attribute. For example, a person assigned a weight of 3,000 lb is likely the result of a typing error made during data collection. This categorization is also essential when selecting the appropriate data analysis or data mining approach to use.

In addition to addressing the mistakes or inconsistencies in data collection, it may be important to change the data to make it more amenable for data analysis. The transformations should be done without losing important information. For example, if a data mining approach requires that all attributes have a consistent range, the data will need to be appropriately modified. The data may also need to be divided into subsets or filtered based on specific criteria to make it amenable to answering the problems outlined at the beginning of the project. Multiple approaches to understanding and preparing data are discussed in Chapters 2 and 3.

1.3.4 Analysis

As discussed earlier, an initial examination of the data is important in understanding the type of information that has been collected and the meaning of the data. In combination with information from the problem definition, this categorization will determine the type of data analysis and data mining approaches to use. Figure 1.4 summarizes some of the main analysis approaches to consider.

Analysis

- Summarizing data
- Exploring relationships between attributes
- Grouping the data
- Identifying non-trivial facts, patterns, and trends
- Building regression models
- Building classification models

FIGURE 1.4   Summary of tasks to consider when analyzing the data.

One common category of analysis tasks provides summarizations and statements about the data. *Summarization* is a process by which data is reduced for interpretation without sacrificing important information. Summaries can be developed for the data as a whole or in part. For example, a retail company that collected data on its transactions could develop summaries of the total sales transactions. In addition, the company could also generate summaries of transactions by products or stores. It may be important to make statements with measures of confidence about the entire data set or groups within the data. For example, if you wish to make a statement concerning the performance of a particular store with slightly lower net revenue than other stores it is being compared to, you need to know if it is really underperforming or just within an expected range of performance. Data visualization, such as charts and summary tables, is an important tool used alongside summarization methods to present broad conclusions and make statements about the data with measures of confidence. These are discussed in Chapters 2 and 4.

A second category of tasks focuses on the identification of important facts, relationships, anomalies, or trends in the data. Discovering this information often involves looking at the data in many ways using a combination of data visualization, data analysis, and data mining methods. For example, a retail company may want to understand customer profiles and other facts that lead to the purchase of certain product lines. *Clustering* is a data analysis method used to group together items with similar attributes. This approach is outlined in Chapter 5. Other data mining methods, such as *decision trees* or *association rules* (also described in Chapter 5), automatically extract important facts or rules from the data. These data mining approaches—describing, looking for relationships, and grouping—combined with data visualization provide the foundation for basic exploratory analysis.

A third category of tasks involves the development of mathematical models that encode relationships in the data. These models are useful for gaining an understanding of the data and for making predictions. To illustrate, suppose a retail company wants to predict whether specific consumers may be interested in buying a particular product. One approach to this problem is to collect historical data containing different customer attributes, such as the customer's age, gender, the location where they live, and so on, as well as which products the customer has purchased in the past. Using these attributes, a mathematical model can be built that encodes important relationships in the data. It may be that female customers between 20 and 35 that live in specific areas are more likely to buy the product. Since these relationships are described in the model, it can be used to examine a list of prospective customers that also contain information on age, gender, location, and so on, to make predictions of those most likely to buy the product. The individuals predicted by the model as buyers of the product might become the focus of a targeted marketing campaign. Models can be built to predict continuous data values (*regression models*) or categorical data (*classification models*). Simple methods to generate these models include *linear regression, logistic regression, classification and regression trees*, and *k-nearest neighbors*. These techniques are discussed in Chapter 6 along with summaries of other approaches. The selection of the methods is often driven by the type of data being analyzed as well as the problem being solved. Some approaches generate solutions that are straightforward to interpret and explain which may be important for examining specific problems. Others are more of a "black box" with limited capabilities for explaining the results. Building and optimizing these models in order to develop useful, simple, and meaningful models can be time-consuming.

There is a great deal of interplay between these three categories of tasks. For example, it is important to summarize the data before building models or finding hidden relationships. Understanding hidden relationships between different items in the data can be of help in generating models. Therefore, it is essential that data analysis or data mining experts work closely with the subject matter expertise in analyzing the data.

1.3.5 Deployment

In the deployment step, analysis is translated into a benefit to the organization and hence this step should be carefully planned and executed. There are many ways to deploy the results of a data analysis or data mining project, as illustrated in Figure 1.5. One option is to write a report for management or the "customer" of the analysis describing the business or scientific intelligence derived from the analysis. The report should be directed to those responsible for making decisions and focused on significant and actionable items—conclusions that can be translated into a decision that can be used to make a difference. It is increasingly common for the report to be delivered through the corporate intranet.

Deployment

- Generate report
- Deploy standalone or integrated decision-support tool
- Measure business impact

FIGURE 1.5 Summary of deployment options.

When the results of the project include the generation of predictive models to use on an ongoing basis, these models can be deployed as standalone applications or integrated with other software such as spreadsheet applications or web services. The integration of the results into existing operational systems or databases is often one of the most cost-effective approaches to delivering a solution. For example, when a sales team requires the results of a predictive model that ranks potential customers based on the likelihood that they will buy a particular product, the model may be integrated with the customer relationship management (CRM) system that they already use on a daily basis. This minimizes the need for training and makes the deployment of results easier. Prediction models or data mining results can also be integrated into systems accessible by your customers, such as e-commerce websites. In the web pages of these sites, additional products or services that may be of interest to the customer may have been identified using a mathematical model embedded in the web server.

Models may also be integrated into existing operational processes where a model needs to be constantly applied to operational data. For example, a solution may detect events leading to errors in a manufacturing system. Catching these errors early may allow a technician to rectify the problem without stopping the production system.

It is important to determine if the findings or generated models are being used to achieve the business objectives outlined at the start of the project. Sometimes the generated models may be functioning as expected but the solution is not being used by the target user community for one reason or another. To increase confidence in the output of the system, a controlled experiment (ideally double-blind) in the field may be undertaken to assess the quality of the results and the organizational impact. For example, the intended users of a predictive model could be divided into two groups. One group, made up of half of the users (randomly selected), uses the model results; the other group does not. The business impact resulting from the two groups can then be measured. Where models are continually updated, the consistency of the results generated should also be monitored over time.

There are a number of deployment issues that may need to be considered during the implementation phase. A solution may involve changing business processes. For example, a solution that requires the development of predictive models to be used by end users in the field may change the work practices of these individuals. The users may even resist this change. A successful method for promoting acceptance is to involve the end users in the definition of the solution, since they will be more inclined to use a system they have helped design. In addition, in order to understand

and trust the results, the users may require that all results be appropriately explained and linked to the data from which the results were generated.

At the end of a project it is always a useful exercise to look back at what worked and what did not work. This will provide insight for improving future projects.

1.4 OVERVIEW OF BOOK

This book outlines a series of introductory methods and approaches important to many data analysis or data mining projects. It is organized into five technical chapters that focus on describing data, preparing data tables, understanding relationships, understanding groups, and building models, with a hands-on tutorial covered in the appendix.

1.4.1 Describing Data

The type of data collected is one of the factors used in the selection of the type of analysis to be used. The information examined on the individual attributes collected in a data set includes a categorization of the attributes' scale in order to understand whether the field represents discrete elements such as *gender* (i.e., male or female) or numeric properties such as *age* or *temperature*. For numeric properties, examining how the data is distributed is important and includes an understanding of where the values of each attribute are centered and how the data for that attribute is distributed around the central values. Histograms, box plots, and descriptive statistics are useful for understanding characteristics of the individual data attributes. Different approaches to characterizing and summarizing elements of a data table are reviewed in Chapter 2, as well as methods that make statements about or summarize the individual attributes.

1.4.2 Preparing Data Tables

For a given data collection, it is rarely the case that the data can be used directly for analysis. The data may contain errors or may have been collected or combined from multiple sources in an inconsistent manner. Many of these errors will be obvious from an inspection of the summary graphs and statistics as well as an inspection of the data. In addition to cleaning the data, it may be necessary to transform the data into a form more amenable for data analysis. Mapping the data onto new ranges, transforming categorical data (such as different colors) into a numeric form to be used in a mathematical model, as well as other approaches to preparing tabular or nonstructured data prior to analysis are reviewed in Chapter 3.

1.4.3 Understanding Relationships

Understanding the relationships between pairs of attributes across the items in the data is the focus of Chapter 4. For example, based on a collection of observations about the population of different types of birds throughout the year as well as the weather conditions collected for a specific region, does the population of a specific bird increase or decrease as the temperature increases? Or, based on a double-blind clinical study, do patients taking a new medication have an improved outcome? Data visualization, such as scatterplots, histograms, and summary tables play an important role in seeing trends in the data. There are also properties that can be calculated to quantify the different types of relationships. Chapter 4 outlines a number of common approaches to understand the relationship between two attributes in the data.

1.4.4 Understanding Groups

Looking at an entire data set can be overwhelming; however, exploring meaningful subsets of items may provide a more effective means of analyzing the data.

Methods for identifying, labeling, and summarizing collections of items are reviewed in Chapter 5. These groups are often based upon the multiple attributes that describe the members of the group and represent subpopulations of interest. For example, a retail store may wish to group a data set containing information about customers in order to understand the types of customers that purchase items from their store. As another example, an insurance company may want to group claims that are associated with fraudulent or nonfraudulent insurance claims. Three methods of automatically identifying such groups—*clustering*, *association rules*, and *decision trees*—are described in Chapter 5.

1.4.5 Building Models

It is possible to encode trends and relationships across multiple attributes as mathematical models. These models are helpful in understanding relationships in the data and are essential for tasks involving the prediction of items with unknown values. For example, a mathematical model could be built from historical data on the performance of windmills as well as geographical and meteorological data concerning their location, and used to make predictions on potential new sites. Chapter 6 introduces important concepts in terms of selecting an approach to modeling, selecting attributes to include in the models, optimization of the models, as well as methods for assessing the quality and usefulness of the models using data not used to create the model. Various modeling approaches are outlined, including *linear regression, logistic regression, classification and regression trees*, and *k-nearest neighbors*. These are described in Chapter 6.

1.4.6 Exercises

At the conclusion of selected chapters, there are a series of exercises to help in understanding the chapters' material. It should be possible to answer these practical exercises by hand and the process of going through them will support learning the material covered. The answers to the exercises are provided in the book's appendix.

1.4.7 Tutorials

Accompanying the book is a piece of software called *Traceis*, which is freely available from the book's website. In the appendix of the book, a series of data analysis and data mining tutorials are provided that provide practical exercises to support learning the concepts in the book using a series of data sets that are available for download.

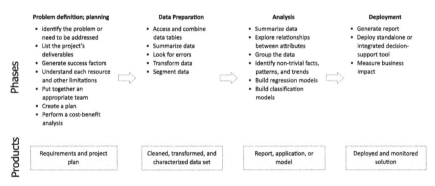

FIGURE 1.6 Summary of steps to consider in developing a data analysis or data mining project.

1.5 SUMMARY

This chapter has described a simple four-step process to use in any data analysis or data mining projects. Figure 1.6 outlines the different stages as well as deliverables to consider when planning and implementing a project to make sense of data.

FURTHER READING

This chapter has reviewed some of the sources of data used in exploratory data analysis and data mining. The following books provide more information on surveys and polls: Fowler (2009), Rea (2005), and Alreck & Settle (2003). There are many additional resources describing experimental design, including Montgomery (2012), Cochran & Cox (1999), Barrentine (1999), and Antony (2003). Operational databases and data warehouses are summarized in the following books: Oppel (2011) and Kimball & Ross (2013). Oppel (2011) also summarizes access and manipulation of information in databases. The CRISP-DM project (CRoss Industry Standard Process for Data Mining) consortium has published in Chapman et al. (2000) a data mining process covering data mining stages and the relationships between the stages. SEMMA (Sample, Explore, Modify, Model, Assess) describes a series of core tasks for model development in the SAS Enterprise Miner[TM] software authored by Rohanizadeh & Moghadam (2009). This chapter has focused on issues relating to large and potentially complex data analysis and data mining projects. There are a number of publications that provide a more detailed treatment of general project management issues, including Berkun (2005), Kerzner (2013), and the Project Management Institute (2013). The following references provide additional case studies: Guidici & Figini (2009), Rud (2000), and Lindoff & Berry (2011).

CHAPTER 2

DESCRIBING DATA

2.1 OVERVIEW

The starting point for data analysis is a data table (often referred to as a data set) which contains the measured or collected data values represented as numbers or text. The data in these tables are called *raw* before they have been transformed or modified. These data values can be measurements of a patient's weight (such as 150 lb, 175 lb, and so on) or they can be different industrial sectors (such as the "telecommunications industry," "energy industry," and so on) used to categorize a company. A data table lists the different items over which the data has been collected or measured, such as different patients or specific companies. In these tables, information considered interesting is shown for different attributes. The individual items are usually shown as rows in a data table and the different attributes shown as columns. This chapter examines ways in which individual attributes can be described and summarized: the scales on which they are measured, how to describe their center as well as the variation using *descriptive statistical* approaches, and how to make statements about these attributes using *inferential statistical* methods, such as confidence intervals or hypothesis tests.

Making Sense of Data I: A Practical Guide to Exploratory Data Analysis and Data Mining,
Second Edition. Glenn J. Myatt and Wayne P. Johnson.
© 2014 John Wiley & Sons, Inc. Published 2014 by John Wiley & Sons, Inc.

2.2 OBSERVATIONS AND VARIABLES

All disciplines collect data about items that are important to that field. Medical researchers collect data on patients, the automotive industry on cars, and retail companies on transactions. These items are organized into a table for data analysis where each row, referred to as an *observation*, contains information about the specific item the row represents. For example, a data table about cars may contain many observations on different types of cars. Data tables also contain information about the car, for example, the car's weight, the number of cylinders, the fuel efficiency, and so on. When an attribute is thought of as a set of values describing some aspect across all observations, it is called a *variable*. An example of a table describing different attributes of cars is shown in Table 2.1 from Bache & Lichman (2013). Each row of the table describes an observation (a specific car) and each column describes a variable (a specific attribute of a car). In this example, there are five observations ("Chevrolet Chevelle Malibu," "Buick Skylark 320," "Plymouth Satellite," "AMC Rebel SST," "Ford Torino") and these observations are described using nine variables: *Name, MPG, Cylinders, Displacement, Horsepower, Weight, Acceleration, Model year,* and *Origin*. (It should be noted that throughout the book variable names in the text will be italicized.)

A generalized version of the data table is shown in Table 2.2, since a table can represent any number of observations described over multiple variables. This table describes a series of observations (from o_1 to o_n) where each observation is described using a series of variables (from x_1 to x_p). A value is provided for each variable of each observation. For example, the value of the first observation for the first variable is x_{11}, the value for the second observation's first variable is x_{21}, and so on. Throughout the book we will explore different mathematical operations that make use of this generalized form of a data table.

The most common way of looking at data is through a spreadsheet, where the raw data is displayed as rows of observations and columns of variables. This type of visualization is helpful in reviewing the raw data; however, the table can be overwhelming when it contains more than a handful of observations or variables. Sorting the table based on one or more variables is useful for organizing the data; however, it is difficult to identify trends or relationships by looking at the raw data alone. An example of a spreadsheet of different cars is shown in Figure 2.1.

Prior to performing data analysis or data mining, it is essential to understand the data table and an important first step is to understand in detail the individual variables. Many data analysis techniques have restrictions on

TABLE 2.1   Data Table Showing Five Car Records Described by Nine Variables

Name	MPG	Cylinders	Displacement	Horsepower	Weight	Acceleration	Model Year	Origin
Chevrolet Chevelle Malibu	18	8	307	130	3504	12	70	America
Buick Skylark 320	15	8	350	165	3693	11.5	70	America
Plymouth Satellite	18	8	318	150	3436	11	70	America
AMC Rebel SST	16	8	304	150	3433	12	70	America
Ford Torino	17	8	302	140	3449	10.5	70	America

TABLE 2.2   Generalized Form of a Data Table

Observations	Variables				
	x_1	x_2	x_3	$\ldots$	x_p
o_1	x_{11}	x_{12}	x_{13}	$\ldots$	x_{1p}
o_2	x_{21}	x_{22}	x_{23}	$\ldots$	x_{2p}
o_3	x_{31}	x_{32}	x_{33}	$\ldots$	x_{3p}
$\ldots$	$\ldots$	$\ldots$	$\ldots$	$\ldots$	$\ldots$
o_n	x_{n1}	x_{n2}	x_{n3}	$\ldots$	x_{np}

the types of variables that they are able to process. As a result, knowing the types of variables allow these techniques to be eliminated from consideration or the data must be transformed into a form appropriate for analysis. In addition, certain characteristics of the variables have implications in terms of how the results of the analysis will be interpreted.

2.3 TYPES OF VARIABLES

Each of the variables within a data table can be examined in different ways. A useful initial categorization is to define each variable based on the type of values the variable has. For example, does the variable contain a fixed number of distinct values (*discrete* variable) or could it take any numeric value (*continuous* variable)? Using the examples from Section 2.1, an *industrial sector* variable whose values can be "telecommunication industry," "retail industry," and so on is an example of a discrete variable since there are a finite number of possible values. A patient's *weight* is an example of a continuous variable since any measured value, such as 153.2 lb, 98.2 lb, is possible within its range. Continuous variables may have an infinite number of values.

car name	mpg	cylinders	displacement	horsepower	weight	acceleration	model year	origin
chevrolet chevelle malibu	18	8	307	130	3504	12	70	American
buick skylark 320	15	8	350	165	3693	11.5	70	American
plymouth satellite	18	8	318	150	3436	11	70	American
amc rebel sst	16	8	304	150	3433	12	70	American
ford torino	17	8	302	140	3449	10.5	70	American
ford galaxie 500	15	8	429	198	4341	10	70	American
chevrolet impala	14	8	454	220	4354	9	70	American
plymouth fury iii	14	8	440	215	4312	8.5	70	American
pontiac catalina	14	8	455	225	4425	10	70	American
amc ambassador dpl	15	8	390	190	3850	8.5	70	American
dodge challenger se	15	8	383	170	3563	10	70	American

FIGURE 2.1   Spreadsheet showing a sample of car observation.

Variables may also be classified according to the *scale* on which they are measured. Scales help us understand the precision of an individual variable and are used to make choices about data visualizations as well as methods of analysis.

A *nominal scale* describes a variable with a limited number of different values that cannot be ordered. For example, a variable *Industry* would be nominal if it had categorical values such as "financial," "engineering," or "retail." Since the values merely assign an observation to a particular category, the order of these values has no meaning.

An *ordinal scale* describes a variable whose values can be ordered or ranked. As with the nominal scale, values are assigned to a fixed number of categories. For example, a scale where the only values are "low," "medium," and "high" tells us that "high" is larger than "medium" and "medium" is larger than "low." However, although the values are ordered, it is impossible to determine the magnitude of the difference between the values. You cannot compare the difference between "high" and "medium" with the difference between "medium" and "low."

An *interval scale* describes values where the interval between values can be compared. For example, when looking at three data values measured on the Fahrenheit scale—5°F, 10°F, 15°F—the differences between the values 5 and 10, and between 10 and 15 are both 5°. Because the intervals between values in the scale share the same unit of measurement, they can be meaningfully compared. However, because the scale lacks a meaningful zero, the ratios of the values cannot be compared. Doubling a value does not imply a doubling of the actual measurement. For example, 10°F is not twice as hot as 5°F.

A *ratio scale* describes variables where both intervals between values and ratios of values can be compared. An example of a ratio scale is a bank account balance whose possible values are $5, $10, and $15. The difference between each pair is $5; and $10 is twice as much as $5. Scales for which it is possible to take ratios of values are defined as having a natural zero.

A variable is referred to as *dichotomous* if it can contain only two values. For example, the values of a variable *Gender* may only be "male" or "female." A *binary* variable is a widely used dichotomous variable with values 0 or 1. For example, a variable *Purchase* may indicate whether a customer bought a particular product using 0 to indicate that a customer did not buy and 1 to indicate that they did buy; or a variable *Fuel Efficiency* may use 0 to represent low efficiency vehicles and 1 to represent high efficiency vehicles. Binary variables are often used in data analysis because they provide a convenient numeric representation for many different types of discrete data and are discussed in detail throughout the book.

Certain types of variables are not used directly in data analysis, but may be helpful for preparing data tables or interpreting the results of the analysis. Sometimes a variable is used to identify each observation in a data table, and will have unique values across the observations. For example, a data table describing different cable television subscribers may include a *customer reference number* variable for each customer. You would never use this variable in data analysis since the values are intended only to provide a link to the individual customers. The analysis of the cable television subscription data may identify a subset of subscribers that are responsible for a disproportionate amount of the company's profit. Including a unique identifier provides a reference to detailed customer information not included in the data table used in the analysis. A variable may also have identical values across the observations. For example, a variable *Calibration* may define the value of an initial setting for a machine used to generate a particular measurement and this value may be the same for all observations. This information, although not used directly in the analysis, is retained both to understand how the data was generated (i.e., what was the calibration setting) and to assess the data for accuracy when it is merged from different sources. In merging data tables generated from two sensors, if the data was generated using different calibration settings then either the two tables cannot be merged or the calibration setting needs to be included to indicate the difference in how the data was measured.

Annotations of variables are another level of detail to consider. They provide important additional information that give insight about the context of the data: Is the variable a count or a fraction? A time or a date? A financial term? A value derived from a mathematical operation on other variables? The units of measurement are useful when presenting the results and are critical for interpreting the data and understanding how the units should align or which transformations apply when data tables are merged from different sources.

In Chapter 6, we further categorize variables (*independent variables* and *response variables*) by the roles they play in the mathematical models generated from data tables.

2.4 CENTRAL TENDENCY

2.4.1 Overview

Of the various ways in which a variable can be summarized, one of the most important is the value used to characterize the center of the set of values it contains. It is useful to quantify the middle or central location of a variable, such as its average, around which many of the observations' values for that variable lie. There are several approaches to calculating this value and which is used can depend on the classification of the variable. The following sections describe some common descriptive statistical approaches for calculating the central location: the *mode*, the *median*, and the *mean*.

2.4.2 Mode

The *mode* is the most commonly reported value for a particular variable. The mode calculation is illustrated using the following variable whose values (after being ordered from low to high) are

$$3, 4, 5, 6, 7, 7, 7, 8, 8, 9$$

The mode would be the value 7 since there are three occurrences of 7 (more than any other value). The mode is a useful indication of the central tendency of a variable, since the most frequently occurring value is often toward the center of the variable's range.

When there is more than one value with the same (and highest) number of occurrences, either all values are reported or a midpoint is selected. For example, for the following values, both 7 and 8 are reported three times:

$$3, 4, 5, 6, 7, 7, 7, 8, 8, 8, 9$$

The mode may be reported as {7, 8} or 7.5.

Mode provides the only measure of central tendency for variables measured on a nominal scale; however, the mode can also be calculated for variables measured on the ordinal, interval, and ratio scales.

2.4.3 Median

The *median* is the middle value of a variable, once it has been sorted from low to high. The following set of values for a variable will be used to illustrate:

$$3, 4, 7, 2, 3, 7, 4, 2, 4, 7, 4$$

Before identifying the median, the values must be sorted:

$$2, 2, 3, 3, 4, 4, 4, 4, 7, 7, 7$$

There are 11 values and therefore the sixth value (five values above and five values below) is selected as the median value, which is 4:

$$2, 2, 3, 3, 4, 4, 4, 4, 7, 7, 7$$

For variables with an even number of values, the average of the two values closest to the middle is selected (sum the two values and divide by 2).

The median can be calculated for variables measured on the ordinal, interval, and ratio scales and is often the best indication of central tendency for variables measured on the ordinal scale. It is also a good indication of the central value for a variable measured on the interval or ratio scales since, unlike the mean, it will not be distorted by extreme values.

2.4.4 Mean

The *mean*—commonly referred to as the average—is the most commonly used summary of central tendency for variables measured on the interval or ratio scales. It is defined as the sum of all the values divided by the number of values. For example, for the following set of values:

$$3, 4, 5, 7, 7, 8, 9, 9, 9$$

The sum of all nine values is $(3 + 4 + 5 + 7 + 7 + 8 + 9 + 9 + 9)$ or 61. The sum divided by the number of values is $61 \div 9$ or 6.78.

For a variable representing a subset of all possible observations (x), the mean is commonly referred to as $\bar{x}$. The formula for calculating a mean, where n is the number of observations and x_i is the individual values, is usually written:

$$\bar{x} = \frac{\sum_{i=1}^{n} x_i}{n}$$

The notation $\sum_{i=1}^{n}$ is used to describe the operation of summing all values of x from the first value $(i = 1)$ to the last value $(i = n)$, that is $x_1 + x_2 + \cdots + x_n$.

2.5 DISTRIBUTION OF THE DATA

2.5.1 Overview

While the central location is a single value that characterizes an individual variable's data values, it provides no insight into the variation of the data or, in other words, how the different values are distributed around this location. The frequency distribution, which is based on a simple count of how many times a value occurs, is often a starting point for the analysis of variation. Understanding the frequency distribution is the focus of the following section and can be performed using simple data visualizations and calculated metrics. As you will see later, the frequency distribution also plays a role in selecting which data analysis approaches to adopt.

2.5.2 Bar Charts and Frequency Histograms

Visualization is an aid to understanding the distribution of data: the range of values, the shape created when the values are plotted, and the values called *outliers* that are found by themselves at the extremes of the range of values. A handful of charts can help to understand the frequency distribution of an individual variable. For a variable measured on a nominal scale, a *bar chart* can be used to display the relative frequencies for the different values. To illustrate, the *Origin* variable from the auto-MPG data table (partially shown in Table 2.2) has three possible values: "America," "Europe," and "Asia." The first step is to count the number of observations in the data table corresponding to each of these values. Out of the 393 observations in the data table, there are 244 observations where the *Origin* is "America," 79 where it is "Asia," and 70 where it is "Europe." In a bar chart, each bar represents a value and the height of the bars is proportional to the frequency, as shown in Figure 2.2.

For nominal variables, the ordering of the x-axis is arbitrary; however, they are often ordered alphabetically or based on the frequency value. The

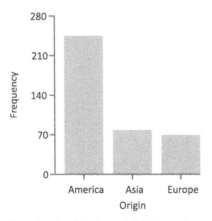

FIGURE 2.2   Bar chart for the *Origin* variable from the auto-MPG data table.

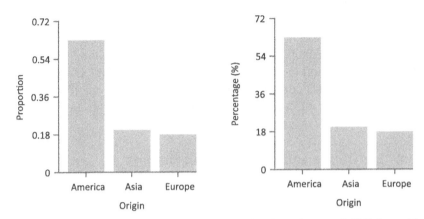

FIGURE 2.3 Bar charts for the *Origin* variables from the auto-MPG data table showing the proportion and percentage.

y-axis which measures frequency can also be replaced by values representing the proportion or percentage of the overall number of observations (replacing the frequency value), as shown in Figure 2.3.

For variables measured on an ordinal scale containing a small number of values, a bar chart can also be used to understand the relative frequencies of the different values. Figure 2.4 shows a bar chart for the variable *PLT* (number of mother's previous premature labors) where there are four possible values: 1, 2, 3, and 4. The bar chart represents the number of values for each of these categories. In this example you can see that most of the observations fall into the "1" category with smaller numbers in the other categories. You can also see that the number of observations decreases as the values increase.

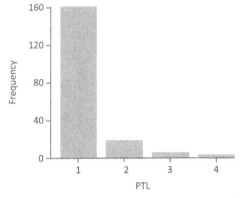

FIGURE 2.4 Bar chart for a variable measured on an ordinal scale, *PLT*.

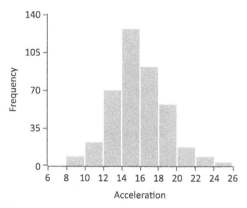

FIGURE 2.5   Frequency histogram for the variable "Acceleration."

The frequency histogram is useful for variables with an ordered scale—ordinal, interval, or ratio—that contain a larger number of values. As with the bar chart, each variable is divided into a series of groups based on the data values and displayed as bars whose heights are proportional to the number of observations within each group. However, the criteria for inclusion within a single bar is a specific range of values. To illustrate, a frequency histogram is shown in Figure 2.5 displaying a frequency distribution for a variable *Acceleration*. The variable has been grouped into a series of ranges from 6 to 8, 8 to 10, 10 to 12, and so on. Since we will need to assign observations that fall on the range boundaries to only one category, we will assign a value to a group where its value is greater than or equal to the lower extreme and less than the upper extreme. For example, an *Acceleration* value of 10 will be categorized into the range 10–12. The number of observations that fall within each range is then determined. In this case, there are six observations that fall into the range 6–8, 22 observations that fall into the range 8–10, and so on. The ranges are ordered from low to high and plotted along the *x*-axis. The height of each histogram bar corresponds to the number of observations for each of the ranges. The histogram in Figure 2.5 indicates that the majority of the observations are grouped in the middle of the distribution between 12 and 20 and there are relatively fewer observations at the extreme values. It is usual to display between 5 and 10 groups in a frequency histogram using boundary values that are easy to interpret.

The frequency histogram helps to understand the shape of the frequency distribution. Figure 2.6 illustrates a number of commonly encountered frequency distributions. The first histogram illustrates a variable where, as the values increase, the number of observations in each group remains

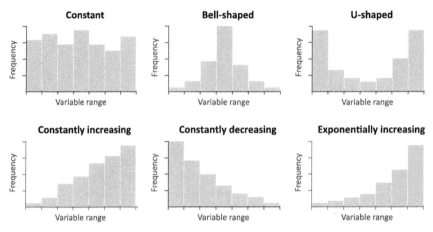

FIGURE 2.6 Examples of frequency distributions.

constant. The second histogram is of a distribution where most of the observations are centered around the mean value, with far fewer observations at the extremes, and with the distribution tapering off toward the extremes. The symmetrical shape of this distribution is often identified as a bell shape and described as a *normal* distribution. It is very common for variables to have a normal distribution and many data analysis techniques assume an approximate normal distribution. The third example depicts a *bimodal* distribution where the values cluster in two locations, in this case primarily at both ends of the distribution. The final three histograms show frequency distributions that either increase or decrease linearly as the values increase (fourth and fifth histogram) or have a nonlinear distribution as in the case of the sixth histogram where the number of observations is increasing exponentially as the values increase.

A frequency histogram can also tell us if there is something unusual about the variables. In Figure 2.7, the first histogram appears to contain two approximately normal distributions and leads us to question whether the data table contains two distinct types of observations, each with a separate

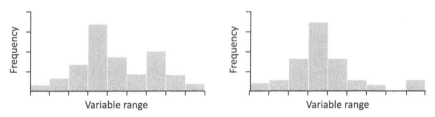

FIGURE 2.7 More complex frequency distributions.

frequency distribution. In the second histogram, there appears to be a small number of high values that do not follow the bell-shaped distribution that the majority of observations follow. In this case, it is possible that these values are errors and need to be further investigated.

2.5.3 Range

The range is a simple measure of the variation for a particular variable. It is calculated as the difference between the highest and lowest values. The following variable will be used to illustrate:

$$2, 3, 4, 6, 7, 7, 8, 9$$

The range is 7 calculated from the highest value (9) minus the lowest value (2). Ranges can be used with variables measured on an ordinal, interval, or ratio scale.

2.5.4 Quartiles

Quartiles divide a continuous variable into four even segments based on the number of observations. The first quartile (Q1) is at the 25% mark, the second quartile (Q2) is at the 50% mark, and the third quartile (Q3) is at the 75% mark. The calculation for Q2 is the same as the median value (described earlier). The following list of values is used to illustrate how quartiles are calculated:

$$3, 4, 7, 2, 3, 7, 4, 2, 4, 7, 4$$

The values are initially sorted:

$$2, 2, 3, 3, 4, 4, 4, 4, 7, 7, 7$$

Next, the median or Q2 is located in the center:

$$2, 2, 3, 3, 4, \mathbf{4}, 4, 4, 7, 7, 7$$

We now look for the center of the first half (shown underlined) or Q1:

$$\underline{2, 2, \mathbf{3}, 3, 4,} \mathbf{4}, 4, 4, 7, 7, 7$$

The value of Q1 is recorded as 3.

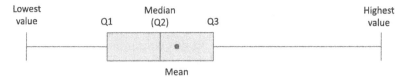

FIGURE 2.8   Overview of elements of a box plot.

Finally, we look for the center of the second half (shown underlined) or Q3:

$$2, 2, 3, 3, 4, \mathbf{4}, \underline{4, 4, \mathbf{7}, 7, 7}$$

The value of Q3 is identified as 7.

When the boundaries of the quartiles do not fall on a specific value, the quartile value is calculated based on the two numbers adjacent to the boundary. The *interquartile range* is defined as the range from Q1 to Q3. In this example it would be $7 - 3$ or 4.

2.5.5 Box Plots

Box plots provide a succinct summary of the overall frequency distribution of a variable. Six values are usually displayed: the lowest value, the lower quartile (Q1), the median (Q2), the upper quartile (Q3), the highest value, and the mean. In the conventional box plot displayed in Figure 2.8, the box in the middle of the plot represents where the central 50% of observations lie. A vertical line shows the location of the median value and a dot represents the location of the mean value. The horizontal line with a vertical stroke between "lowest value" and "Q1" and "Q3" and "highest value" are the "tails"—the values in the first and fourth quartiles.

Figure 2.9 provides an example of a box plot for one variable (*MPG*). The plot visually displays the lower (9) and upper (46.6) bounds of the variable. Fifty percent of observations begin at the lower quartile (17.5)

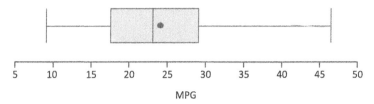

FIGURE 2.9   Box plot for the variable *MPG*.

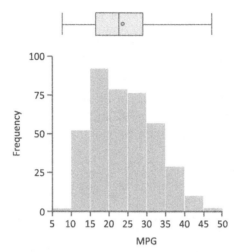

FIGURE 2.10 Comparison of frequency histogram and a box plot for the variable *MPG*.

and end at the upper quartile (29). The median and the mean values are close, with the mean slightly higher (around 23.6) than the median (23). Figure 2.10 shows a box plot and a histogram side-by-side to illustrate how the distribution of a variable is summarized using the box plot.

"Outliers," the solitary data values close to the ends of the range of values, are treated differently in various forms of the box plot. Some box plots do not graphically separate them from the first and fourth quartile depicted by the horizontal lines that are to the left and the right of the box. In other forms of box plots, these extreme values are replaced with the highest and lowest values not considered an outlier and the outliers are explicitly drawn (using small circles) outside the main plot as shown in Figure 2.11.

Box plots help in understanding the symmetry of a frequency distribution. If both the mean and median have approximately the same value,

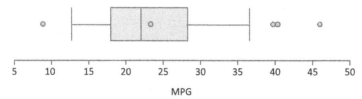

FIGURE 2.11 A box plot with extreme values explicitly shown as circles.

there will be about the same number of values above and below the mean and the distribution will be roughly symmetric.

2.5.6 Variance

The *variance* describes the spread of the data and measures how much the values of a variable differ from the mean. For variables that represent only a sample of some population and not the population as a whole, the variance formula is

$$s^2 = \frac{\sum\limits_{i=1}^{n}(x_i - \bar{x})^2}{n - 1}$$

The sample variance is referred to as s^2. The actual value (x_i) minus the mean value ($\bar{x}$) is squared and summed for all values of a variable. This value is divided by the number of observations minus 1 ($n - 1$).

The following example illustrates the calculation of a variance for a particular variable:

$$3, 4, 4, 5, 5, 5, 6, 6, 6, 7, 7, 8, 9$$

where the mean is

$$\bar{x} = \frac{3 + 4 + 4 + 5 + 5 + 5 + 6 + 6 + 6 + 7 + 7 + 8 + 9}{13}$$

$$\bar{x} = 5.8$$

Table 2.3 is used to calculate the sum, using the mean value of 5.8.

To calculate s^2, we substitute the values from Table 2.3 into the variance formula:

$$s^2 = \frac{\sum\limits_{i=1}^{n}(x_i - \bar{x})^2}{n - 1}$$

$$s^2 = \frac{34.32}{13 - 1}$$

$$s^2 = 2.86$$

The variance reflects the average squared deviation and can be calculated for variables measured on the interval or ratio scale.

TABLE 2.3   Variance Intermediate Steps

x	$\bar{x}$	$(x_i - \bar{x})$	$(x_i - \bar{x})^2$
3	5.8	−2.8	7.84
4	5.8	−1.8	3.24
4	5.8	−1.8	3.24
5	5.8	−0.8	0.64
5	5.8	−0.8	0.64
5	5.8	−0.8	0.64
6	5.8	0.2	0.04
6	5.8	0.2	0.04
6	5.8	0.2	0.04
7	5.8	1.2	1.44
7	5.8	1.2	1.44
8	5.8	2.2	4.84
9	5.8	3.2	10.24
			Sum = 34.32

2.5.7 Standard Deviation

The *standard deviation* is the square root of the variance. For a sample from a population, the formula is

$$s = \sqrt{\frac{\sum\limits_{i=1}^{n} (x_i - \bar{x})^2}{n - 1}}$$

where s is the sample standard deviation, x_i is the actual data value, $\bar{x}$ is the mean for the variable, and n is the number of observations. For a calculated variance (e.g., 2.86) the standard deviation is calculated as $\sqrt{2.86}$ or 1.69.

The standard deviation is the most widely used measure of the deviation of a variable. The higher the value, the more widely distributed the variable's data values are around the mean. Assuming the frequency distribution is approximately normal (i.e., a bell-shaped curve), about 68% of all observations will fall within one standard deviation of the mean (34% less than and 34% greater than). For example, a variable has a mean value of 45 with a standard deviation value of 6. Approximately 68% of the observations should be in the range 39–51 (45 ± one standard deviation) and approximately 95% of all observations fall within two standard deviations of the mean (between 33 and 57). Standard deviations can be calculated for variables measured on the interval or ratio scales.

It is possible to calculate a normalized value, called a *z-score*, for each data element that represents the number of standard deviations that element's value is from the mean. The following formula is used to calculate the *z*-score:

$$z = \frac{x_i - \bar{x}}{s}$$

where z is the *z*-score, x_i is the actual data value, $\bar{x}$ is the mean for the variable, and s is the standard deviation. A *z*-score of 0 indicates that a data element's value is the same as the mean, data elements with *z*-scores greater than 0 have values greater than the mean, and elements with *z*-scores less than 0 have values less than the mean. The magnitude of the *z*-score reflects the number of standard deviations that value is from the mean. This calculation can be useful for comparing variables measured on different scales.

2.5.8 Shape

Previously in this chapter, we discussed ways to visualize the frequency distribution. In addition to these visualizations, there are methods for quantifying the lack of symmetry or *skewness* in the distribution of a variable. For asymmetric distributions, the bulk of the observations are either to the left or the right of the mean. For example, in Figure 2.12 the frequency distribution is asymmetric and more of the observations are to the left of the mean than to the right; the right tail is longer than the left tail. This is an example of a positive, or right skew. Similarly, a negative, or left skew would have more of the observations to the right of the mean value with a longer tail on the left.

It is possible to calculate a value for skewness that describes whether the variable is positively or negatively skewed and the degree of skewness. One formula for estimating skewness, where the variable is x with individual values x_i, and n data values is

$$skewness = \left(\frac{\sqrt{n \times (n-1)}}{n-2} \right) \times \frac{\frac{1}{n} \times \sum_{i=1}^{n}(x_i - \bar{x})^3}{\left(\frac{1}{n} \times \sum_{i=1}^{n}(x_i - \bar{x})^2 \right)^{3/2}}$$

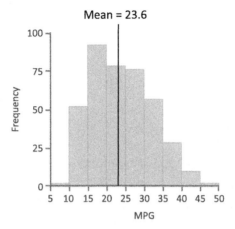

FIGURE 2.12 Frequency distribution showing a positive skew.

A skewness value of zero indicates a symmetric distribution. If the lower tail is longer than the upper tail the value is positive; if the upper tail is longer than the lower tail, the skewness score is negative. Figure 2.13 shows examples of skewness values for two variables. The variable *alkphos* in the plot on the left has a positive skewness value of 0.763, indicating that the majority of observations are to the left of the mean, whereas the negative skewness value for the variable *mcv* in the plot on the right indicates that the majority are to the right of the mean. That the skewness value for *mcv* is closer to zero than *alkphos* indicates that *mcv* is more symmetric than *alkphos*.

In addition to the symmetry of the distribution, the type of peak the distribution has should be considered and it can be characterized by a measurement called *kurtosis*. The following formula can be used for

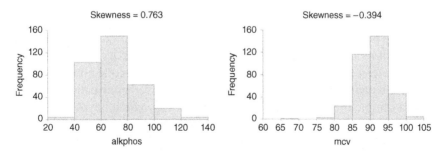

FIGURE 2.13 Skewness estimates for two variables.

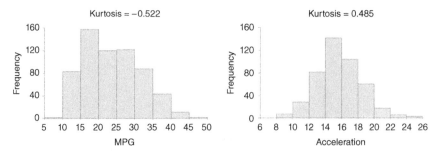

FIGURE 2.14 Kurtosis estimates for two variables.

calculating kurtosis for a variable x, where x_i represents the individual values, and n the number of data values:

$$kurtosis = \frac{n-1}{(n-2)\times(n-3)} \times \left((n+1) \times \frac{\sum_{i=1}^{n}\left(x_i - \bar{x}\right)^4 / n}{\left(\sum_{i=1}^{n}(x_i - x)^2 / n\right)^2} - 3 \right) + 6$$

Variables with a pronounced peak near the mean have a high kurtosis score while variables with a flat peak have a low kurtosis score. Figure 2.14 illustrates kurtosis scores for two variables.

It is important to understand whether a variable has a normal distribution, since a number of data analysis approaches require variables to have this type of frequency distribution. Values for skewness and kurtosis close to zero indicate that the shape of a frequency distribution for a variable approximates a normal distribution which is important for checking assumptions in certain data analysis methods.

2.6 CONFIDENCE INTERVALS

Up to this point, we have been looking at ways of summarizing information on a set of randomly collected observations. This summary information is usually referred to as *statistics* as they summarize only a collection of observations that is a subset of a larger population. However, information derived from a sample of observations can only be an approximation of the entire population. To make a definitive statement about an entire *population*, every member of that population would need to be measured. For example, if we wanted to say for certain that the average weight of men in the United States is 194.7 lb, we would have to collect the weight measurements for every man living in the United States and derive a mean from these observations. This is not possible or practical in most situations.

It is possible, however, to make estimates about a population by using *confidence intervals*. Confidence intervals are a measure of our uncertainty about the statistics we calculate from a single sample of observations. For example, the confidence interval might state that the average weight of men in the United States is between 191.2 lb and 198.2 lb to take into account the uncertainty of measuring only a sample of the total population. Only if the sample of observations is a truly random sample of the entire population can these types of estimates be made.

To understand how a statistic, such as the mean or mode, calculated from a single sample can reliably be used to infer a corresponding value of the population that cannot be measured and is therefore unknown, you need to understand something about *sample distributions*. Each statistic has a corresponding unknown value in the population called a *parameter* that can be estimated. In the example used in this section, we chose to calculate the statistic *mean* for the weight of US males. The mean value for the random sample selected is calculated to be 194.7 lb. If another random sample with the same number of observations were collected, the mean could also be calculated and it is likely that the means of the two samples would be close but not identical. If we take many random samples of equal size and calculate the mean value from each sample, we would begin to form a frequency distribution. If we were to take infinitely many samples of equal size and plot on a graph the value of the mean calculated from each sample, it would produce a *normal frequency distribution* that reflects the distribution of the sample means for the population mean under consideration. The distribution of a statistic computed for each of many random samples is called a *sampling distribution*. In our example, we would call this the *sampling distribution of the mean*.

Just as the distributions for a statistical variable discussed in earlier sections have a mean and a standard deviation, so also does the sampling distribution. However, to make clear when these measures are being used to describe the distribution of a statistic rather than the distribution of a variable, distinct names are used. The mean of a sampling distribution is called the *expected value of the mean*: it is the mean expected of the population. The standard deviation of the sampling distribution is called the *standard error*: it measures how much error to expect from equally sized random samples drawn from the same population. The standard error informs us of the average difference between the mean of a sample and the expected value.

The sample size is important. It is beyond the scope of this book to explain the details, but regardless of how the values of a variable for the population are distributed, the sampling distribution of a statistic calculated on samples from that variable will have a *normal* form when the size chosen for the samples has at least 30 observations. This is known as the *law of large numbers,* or more formally as the *central limit theorem.*

The standard error plays a fundamental role in inferential statistics by providing a measurable level of confidence in how well a sample mean estimates the mean of the population. The standard error can be calculated from a sample using the following formula:

$$standard\ error\ of\ the\ sampling\ distribution = \frac{s}{\sqrt{n}}$$

where s is the standard deviation of a sample and n is the number of observations in the sample. Because the size n is in the denominator and the standard deviation s is in the numerator, small samples with large variations increase the standard error, reducing the confidence that the sample statistic is a close approximation of the population parameter we are trying to estimate.

The data analyst or the team calculating the confidence interval should decide what the desired level of confidence should be. Confidence intervals are often based on a 95% confidence level, but sometimes a more stringent 99% confidence level or less stringent 90% level is used. Using a confidence interval of 95% to illustrate, one way to interpret this confidence level is that, on average, the correct population value will be found within the derived confidence interval 95 times out of every 100 samples collected. In these 100 samples, there will be 5 occasions on average when this value does not fall within the range. The confidence level is usually stated in terms of α from the following equation:

$$confidence\ interval = 100 \times (1 - \alpha)$$

For a 90% confidence level alpha is 0.1; for a 95% confidence level alpha is 0.05; for a 99% confidence level α is 0.01; and so on. The value used for this level of confidence will affect the size of the interval; that is, the higher the desired level of confidence the wider the confidence interval.

Along with the value of α selected, the confidence interval is based on the standard error. The estimated range or confidence interval is calculated using this confidence level along with information on the number of

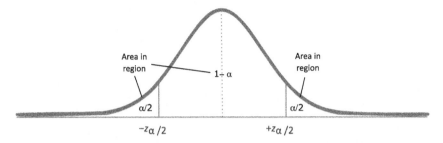

FIGURE 2.15   Illustration of the standard z-distribution to calculate $z_{\alpha/2}$.

observations in the sample as well as the variation in the sample's data. The formula showing a confidence interval for a mean value is shown here:

$$\bar{x} \pm z_{\alpha/2} \left(\frac{s}{\sqrt{n}} \right)$$

where $\bar{x}$ is the mean value, s is the standard deviation, and n is the number of observations in the sample. The value for $z_{\alpha/2}$ is based on the area to the right of a standard z-distribution as illustrated in Figure 2.15 since the total area under this curve is 1. This number can be derived from a standard statistical table or computer program.

To illustrate, the fuel efficiency of 100 specific cars is measured and a mean value of 30.35 *MPG* is calculated with a standard deviation of 2.01. Using an alpha value of 0.05 (which translates into a $z_{\alpha/2}$ value of 1.96), the confidence interval is calculated as

$$\bar{x} \pm z_{\alpha/2} \left(\frac{s}{\sqrt{n}} \right)$$

$$30.35 \pm 1.96 \left(\frac{2.01}{\sqrt{100}} \right)$$

Hence, the confidence interval for the average fuel efficiency is 30.35 ± 0.393 or between 29.957 and 30.743.

For calculation of a confidence interval where sigma is unknown and the number of observations is less than 30 observations, a t-distribution should be used (see Urdan (2010), Anderson et al. (2010), Witte & Witte (2009), Kachigan (1991), Freedman et al. (2007), and Vickers (2010) for more details).

2.7 HYPOTHESIS TESTS

Hypothesis tests are used to support making decisions by helping to understand whether data collected from a sample of all possible observations supports a particular hypothesis. For example, a company manufacturing hair care products wishes to say that the average amount of shampoo within the bottles is 200 mL. To test this hypothesis, the company collects a random sample of 100 shampoo bottles and precisely measures the contents of the bottle. If it is inferred from the sample that the average amount of shampoo in each bottle is not 200 mL then a decision may be made to stop production and rectify the manufacturing problem.

The first step is to formulate the hypothesis that will be tested. This hypothesis is referred to as the null hypothesis (H_0). The null hypothesis is stated in terms of what would be expected if there were nothing unusual about the measured values of the observations in the data from the samples we collect—"null" implies the absence of effect. In the example above, if we expected each bottle of shampoo to contain 200 mL of shampoo, the null hypothesis would be: the average volume of shampoo in a bottle is 200 mL. Its corresponding alternative hypothesis (H_a) is that they differ or, stated in a way that can be measured, that the average is not equal to 200 mL. For this example, the null hypothesis and alternative hypothesis would be shown as

$$H_0 : \mu = 200$$
$$H_a : \mu \neq 200$$

This hypothesis will be tested using the sample data collected to determine whether the mean value is different enough to warrant rejecting the null hypothesis. Hence, the result of a hypothesis test is either to *fail to reject* or *reject* the null hypothesis. Since we are only looking at a sample of the observations—we are not testing every bottle being manufactured—it is impossible to make a statement about the average with total certainty. Consequently, it is possible to reach an incorrect conclusion. There are two categories of errors that can be made. One is to reject the null hypothesis when, in fact, the null hypothesis should stand (referred to as a type I error); the other is to accept the null hypothesis when it should be rejected (or type II error). The threshold of probability used to determine a type I error should be decided before performing the test. This threshold, which is also referred to as the level of significance or α, is often set to 0.05 (5% chance of a type I error); however, more stringent (such as 0.01 or 1%

chance) or less stringent values (such as 0.1 or 10% chance) can be used depending on the consequences of an incorrect decision.

The next step is to specify the standardized test statistic (T). We are interested in determining whether the average of the sample data we collected is either meaningfully or trivially different from the population average. Is it likely that we would find as great a difference from the population average were we to collect other random samples of the same size and compare their mean values? Because the hypothesis involves the mean, we use the following formula to calculate the test statistic:

$$T = \frac{\bar{x} - \mu_0}{s / \sqrt{n}}$$

where $\bar{x}$ is the calculated mean value of the sample, μ_0 is the population mean that is the subject of the hypothesis test, s is the standard deviation of the sample, and n is the number of observations. (Recall that the denominator is the standard error of the sampling distribution.) In this example, the average shampoo bottle volume measured over the 100 samples (n) is 199.94 ($\bar{x}$) and the standard deviation is 0.613 (s).

$$T = \frac{199.94 - 200}{0.613 / \sqrt{100}} = -0.979$$

Assuming we are using a value for α of 0.05 as the significance level to formulate a decision rule to either let the null hypothesis stand or reject it, it is necessary to identify a range of values where 95% of all the sample means would lie. As discussed in Section 2.6, the law of large numbers applies to the sampling distribution of the statistic T: when there are at least 30 observations, the frequency distribution of the sample means is approximately normal and we can use this distribution to estimate regions to accept the null hypothesis. This region has two critical upper and lower bound values C1 and C2. Ninety-five percent of all sample means lie between these values and 5% lie outside these values (0.025 below C2 and 0.025 above C1) (see Figure 2.16). We reject the null hypothesis if the value of T is outside this range (i.e., greater than C1 or less than C2) or let the null hypothesis stand if it is inside the range.

Values for C1 and C2 can be calculated using a standard z-distribution table lookup and would be C2 = -1.96 and C1 = $+1.96$. These z-values were selected where the combined area to the left of C2 and to the right of C1 would equal 0.05.

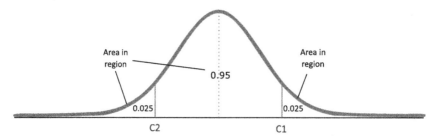

FIGURE 2.16   Standard z-distribution.

Since T is -0.979 and is greater than -1.96 and less than $+1.96$, we let the null hypothesis stand and conclude that the value is within the acceptable range.

In this example, the hypothesis test is referred to as a two-tailed test—that is, we tested the hypothesis for values above and below the critical values; however, hypothesis tests can be structured such that they are testing for values only above or below a value.

It is a standard practice to also calculate a *p-value* which corresponds to the probability of obtaining a test statistic value at least as extreme as the observed value (assuming the null hypothesis is true). This p-value can also be used to assess the null hypothesis, where the null hypothesis is rejected if it is less than the value of alpha. This value can be looked up using a standard z-distribution table as found by an online search or readily available software. In this example, the p-value would be 0.33. Since this value is not less than 0.05 (as defined earlier) we again do not reject the null hypothesis.

EXERCISES

A set of 10 hypothetical patient records from a large database is presented in Table 2.4. Patients with a diabetes value of 1 have type-II diabetes and patients with a diabetes value of 0 do not have type-II diabetes.

1. For each of the following variables, assign them to one of the following scales: nominal, ordinal, interval, or ratio:
   (a) *Name*
   (b) *Age*
   (c) *Gender*
   (d) *Blood group*
   (e) *Weight (kg)*

TABLE 2.4 Table of Patient Records

Name	Age	Gender	Blood Group	Weight (kg)	Height (m)	Systolic Blood Pressure (mmHg)	Diastolic Blood Pressure (mmHg)	Diabetes
P. Lee	35	Female	A Rh+	50	1.52	68	112	0
R. Jones	52	Male	O Rh−	115	1.77	110	154	1
J. Smith	45	Male	O Rh+	96	1.83	88	136	0
A. Patel	70	Female	O Rh−	41	1.55	76	125	0
M. Owen	24	Male	A Rh−	79	1.82	65	105	0
S. Green	43	Male	O Rh−	109	1.89	114	159	1
N. Cook	68	Male	A Rh+	73	1.76	108	136	0
W. Hands	77	Female	O Rh−	104	1.71	107	145	1
P. Rice	45	Female	O Rh+	64	1.74	101	132	0
F. Marsh	28	Male	O Rh+	136	1.78	121	165	1

(f) *Height (m)*

(g) *Systolic blood pressure (mmHg)*

(h) *Diastolic blood pressure (mmHg)*

(i) *Diabetes*

TABLE 2.5 Table with Variables *Name* and *Age*

Name	Age
P. Lee	35
R. Jones	52
J. Smith	45
A. Patel	70
M. Owen	24
S. Green	43
N. Cook	68
W. Hands	77
P. Rice	45
F. Marsh	28

TABLE 2.6 Retail Transaction Data Set

Customer	Store	Product Category	Product Description	Sale Price ($)	Profit ($)
B. March	New York, NY	Laptop	DR2984	950	190
B. March	New York, NY	Printer	FW288	350	105
B. March	New York, NY	Scanner	BW9338	400	100
J. Bain	New York, NY	Scanner	BW9443	500	125
T. Goss	Washington, DC	Printer	FW199	200	60
T. Goss	Washington, DC	Scanner	BW39339	550	140
L. Nye	New York, NY	Desktop	LR21	600	60
L. Nye	New York, NY	Printer	FW299	300	90
S. Cann	Washington, DC	Desktop	LR21	600	60
E. Sims	Washington, DC	Laptop	DR2983	700	140
P. Judd	New York, NY	Desktop	LR22	700	70
P. Judd	New York, NY	Scanner	FJ3999	200	50
G. Hinton	Washington, DC	Laptop	DR2983	700	140
G. Hinton	Washington, DC	Desktop	LR21	600	60
G. Hinton	Washington, DC	Printer	FW288	350	105
G. Hinton	Washington, DC	Scanner	BW9443	500	125
H. Fu	New York, NY	Desktop	ZX88	450	45
H. Taylor	New York, NY	Scanner	BW9338	400	100

2. Calculate the following statistics for the variable *Age* (from Table 2.5):

(**a**) Mode

(**b**) Median

(**c**) Mean

(**d**) Range

(**e**) Variance

(**f**) Standard deviation

3. Using the data in Table 2.6, create a histogram of *Sale Price ($)* using the following intervals: 0 to less than 250, 250 to less than 500, 500 to less than 750, and 750 to less than 1000.

FURTHER READING

A number of books provide basic introductions to statistical methods including Donnelly (2007) and Levine & Stephan (2010). Numerous books provide additional details on the descriptive and inferential statistics as, for example, Urdan (2010), Anderson et al. (2010), Witte & Witte (2009), Kachigan (1991), Freedman et al. (2007), and Vickers (2010). The conceptual difference between standard error and standard deviation described in Sections 2.6 and 2.7 is often difficult to grasp. For further discussion, see the section on sampling distributions in Kachigan (1991) and the chapter on standard error in Vickers (2010). For further reading on communicating information, see Tufte (1990, 1997a, 1997b, 2001, 2006). These works describe a theory of data graphics and information visualization that are illustrated by many examples.

CHAPTER 3

PREPARING DATA TABLES

3.1 OVERVIEW

Preparing the data is one of the most time-consuming parts of a data analysis/data mining project. This chapter outlines concepts and steps necessary to prepare a data set prior to beginning data analysis or data mining. The way in which the data is collected and prepared is critical to the confidence with which decisions can be made. The data needs to be merged into a table and this may involve integration of the data from multiple sources. Once the data is in a tabular format, it should be fully characterized as discussed in the previous chapter. The data should be cleaned by resolving ambiguities and errors, removing redundant and problematic data, and eliminating columns of data irrelevant to the analysis. New columns of data may need to be calculated. Finally, the table should be divided, where appropriate, into subsets that either simplify the analysis or allow specific questions to be answered more easily.

In addition to the work done preparing the data, it is important to record the details about the steps that were taken and why they were done. This not only provides documentation of the activities performed so far, but it also provides a methodology to apply to similar data sets in the future. In

Making Sense of Data I: A Practical Guide to Exploratory Data Analysis and Data Mining,
Second Edition. Glenn J. Myatt and Wayne P. Johnson.
© 2014 John Wiley & Sons, Inc. Published 2014 by John Wiley & Sons, Inc.

addition, when validating the results, these records will be important for recalling assumptions made about the data.

The following chapter outlines the process of preparing data for analysis. It includes methods for identifying and cleaning up errors, removing certain variables or observations, generating consistent scales across different observations, generating new frequency distributions, converting text to numbers and vice versa, combining variables, generating groups, and preparing unstructured data.

3.2 CLEANING THE DATA

For variables measured on a nominal or ordinal scale (where there are a fixed number of possible values) it is useful to inspect all possible values to uncover mistakes, duplications and inconsistencies. Each value should map onto a unique term. For example, a variable *Company* may include a number of different spellings for the same company such as "General Electric Company," "General Elec. Co.," "GE," "Gen. Electric Company," "General electric company," and "G.E. Company." When these values refer to the same company, the various terms should be consolidated into one. Subject matter expertise may be needed to correct and harmonize these variables. For example, a company name may include one of the divisions of the General Electric Company and for the purpose of this specific project it should also be included as the "General Electric Company."

A common problem with numeric variables is the inclusion of non-numeric terms. For example, a variable generally consisting of numbers may include a value such as "above 50" or "out of range." Numeric analysis cannot interpret a non-numeric value and hence, relying on subject matter expertise, these terms should be converted to a number or the observation removed.

Another problem arises when observations for a particular variable are missing data values. Where there is a specific meaning for a missing data value, the value may be replaced based on knowledge of how the data was collected.

It can be more challenging to clean variables measured on an interval or ratio scale since they can take any possible value within a range. However, it is useful to consider outliers in the data. Outliers are a single or a small number of data values that differ greatly from the rest of the values. There are many reasons for outliers. For example, an outlier may be an error in the measurement or the result of measurements made using a different calibration. An outlier may also be a legitimate and valuable data point.

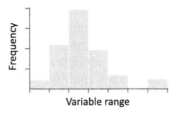

FIGURE 3.1   Histogram showing an outlier.

Histograms and box plots can be useful in identifying outliers as previously described. The histogram in Figure 3.1 displays a variable *Height* where one value is eight times higher than the average of all data points.

A particular variable may have been measured over different units. For example, a variable *Weight* may have been measured using both pounds and kilograms for different observations or a variable *Price* may be measured in different currencies. These should be standardized to a single scale so that they can be compared during analysis. In situations where data has been collected over time, changes related to the passing of time may no longer be relevant to the analysis. For example, when looking at a variable *Cost of production* for which the data has been collected over many years, the rise in costs attributable to inflation may need to be considered for the analysis. When data is combined from multiple sources, an observation is more likely to have been recorded more than once. Duplicate entries should be removed.

3.3 REMOVING OBSERVATIONS AND VARIABLES

After an initial categorization of the variables, it may be possible to remove variables from consideration. For example, constants and variables with too many missing data values would be candidates for removal. Similarly, it may be necessary to remove observations that have data missing for a particular variable. For more information on this process, see Section 3.10.

3.4 GENERATING CONSISTENT SCALES ACROSS VARIABLES

Sometimes data analysis and data mining programs have difficulty processing data in its raw form. For these cases, certain mathematical transformations can be applied to the data. *Normalization* uses a mathematical function to transform numeric columns to a new range. Normalization is important in preventing certain data analysis methods from giving some variables undue influence over others because of differences in the range of their values. For example, when analyzing customer credit card data, the *Credit limit* value (whose values might range from $500 to $100,000) should not be given more weight in the analysis than the *Customer's age* (whose values might range from 18 to 100).

The *min–max* transformation maps the values of a variable to a new range, such as from 0 to 1. The following formula is used:

$$x'_i = \frac{x_i - OriginalMin}{OriginalMax - OriginalMin} \times (NewMax - NewMin) + NewMin$$

where x'_i is the new normalized value, x_i is the original variable's value, *OriginalMin* is the minimum possible value in the original variable, *OriginalMax* is the maximum original possible value, *NewMin* is the minimum value for the normalized range, and *NewMax* is the maximum value for the normalized range. Since the minimum and maximum values for the original variable are needed, if the original data does not contain the full range, either an estimate of the range is needed or the formula should be restricted to the range specified for future use.

The *z-score* transformation normalizes the values around the mean of the set, with differences from the mean being recorded as standardized units, based on the frequency distribution of the variable, as discussed in Section 2.5.7.

The *decimal scaling* transformation moves the decimal point to ensure the range is between 1 and −1. The following formula is used:

$$x'_i = \frac{x_i}{10^n}$$

where n is the number of digits of the maximum absolute value. For example, if the largest number is 9948 then n would be 4. 9948 would normalize to $9948/10^4$ or 9948/10,000 or 0.9948.

The normalization process is illustrated using the data in Table 3.1. As an example, to calculate the normalized values using the min–max equation for the variable *Weight*, first the minimum and maximum values should be identified: *OriginalMin* = 1613 and *OriginalMax* = 5140. The new normalized values will be between 0 and 1, hence *NewMin* = 0 and

TABLE 3.1 Normalization of the Variable Weight Using the Min–Max, z-score, and Decimal Scaling Transformations

Car Name	Weight	Min–Max (Weight)	z-score (Weight)	Decimal Scaling (Weight)
Datsun 1200	1613	0	−1.59	0.161
Honda Civic Cvcc	1800	0.053	−1.37	0.18
Volkswagen Rabbit	1825	0.0601	−1.34	0.182
Renault 5 gtl	1825	0.0601	−1.34	0.182
Volkswagen Super Beetle	1950	0.0955	−1.19	0.195
Mazda glc 4	1985	0.105	−1.15	0.198
Ford Pinto	2046	0.123	−1.08	0.205
Plymouth Horizon	2200	0.166	−0.898	0.22
Toyota Corolla	2265	0.185	−0.822	0.226
AMC Spirit dl	2670	0.3	−0.345	0.267
Ford Maverick	3158	0.438	0.229	0.316
Plymouth Volare Premier v8	3940	0.66	1.15	0.394
Dodge d200	4382	0.785	1.67	0.438
Pontiac Safari (sw)	5140	1	2.56	0.514

$NewMax = 1$. To calculate the new min–max normalized value for the Ford maverick using the formula:

$$x_i' = \frac{x_i - OriginalMin}{OriginalMax - OriginalMin} \times (NewMax - NewMin) + NewMin$$

$$x_i' = \frac{3158 - 1613}{5140 - 1613} \times (1 - 0) + 0$$

$$x_i' = 0.438$$

Table 3.1 shows some of the calculated normalized values for the min–max normalization, the z-score normalization, and the decimal scaling normalization.

3.5 NEW FREQUENCY DISTRIBUTION

A variable may not conform to a normal frequency distribution; however, certain data analysis methods may require that the data follow a normal distribution. Methods for visualizing and describing normal frequency distributions are described in the previous chapter. To transform the data

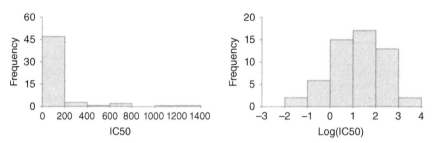

FIGURE 3.2 Log transformation converting a variable (IC50) to adjust the frequency distribution.

so that it more closely approximates a normal distribution, it may be necessary to take the *log*, *exponential*, or a *Box–Cox* transformation. The formula for a *Box–Cox* transformation is

$$x'_i = \frac{x_i^{\lambda} - 1}{\lambda}$$

where λ is a value greater than 1.

Figure 3.2 shows how the distribution of the original variable IC50 (figure on the left) is transformed to a closer approximation of the normal distribution after the log transformation has been applied (figure on the right).

3.6 CONVERTING TEXT TO NUMBERS

To use variables that have been assigned as nominal or ordinal and described using text values within certain numerical analysis methods, it is necessary to convert the variable's values into numbers. For example, a variable with values "low," "medium" and "high" may have "low" replaced by 0, "medium" replaced by 1, and "high" replaced by 2.

Another way to handle nominal data is to convert each value into a separate column with values 1 (indicating the presence of the category) and 0 (indicating the absence of the category). These new variables are often referred to as *dummy* variables. In Table 3.2, for example, the variable *Color* has now been divided into five separate columns, one for each color. While we have shown a column for each color, in practice five dummy variables are not needed to encode the five colors. We could get by with only four variables (*Color* = "Black," *Color* = "Blue," *Color* = "Red," and *Color* = "Green"). To represent the five colors, the values for the "Black,"

TABLE 3.2 Generating a Series of Dummy Variables from the Single *Color* **Variable**

Product ID	Color	Color = Black	Color = Blue	Color = Red	Color = Green	Color = White
89893-22	Black	1	0	0	0	0
849082-35	Blue	0	1	0	0	0
27037-84	Red	0	0	1	0	0
2067-09	Green	0	0	0	1	0
44712-61	White	0	0	0	0	1
98382-34	Blue	0	1	0	0	0
72097-52	Green	0	0	0	1	0

"Blue," "Red," and "Green" variables would be for Black: 1,0,0,0, for Blue: 0,1,0,0, for Red: 0,0,1,0, for Green: 0,0,0,1, and for White: 0,0,0,0.

3.7 CONVERTING CONTINUOUS DATA TO CATEGORIES

By converting continuous data into discrete values, it might appear that we are reducing the information content of the variable. However, this conversion is desirable in a number of situations. First, where a value is defined on an interval or ratio scale but when knowledge about how the data was collected suggests the accuracy of the data does not warrant these scales, a variable may be a candidate for conversion to a categorical variable that reflects the true variation in the data. Second, because certain techniques can only process categorical data, converting continuous data into discrete values makes a numeric variable accessible to these methods. For example, a continuous variable *credit score* may be divided into four categories: poor, average, good, and excellent; or a variable *Weight* that has a range from 0 to 350 lb may be divided into five categories: less than 100 lb, 100–150 lb, 150–200 lb, 200–250 lb, and above 250 lb. All values for the variable *Weight* must now be assigned to a category and assigned an appropriate value such as the mean of the assigned category. It is often useful to use the frequency distribution to understand appropriate range boundaries.

This process can also be applied to nominal variables, especially in situations where there are a large number of values for a given nominal variable. If the data set were to be summarized using each of the values, the number of observations for each value may be too small to reach any meaningful conclusions. However, a new variable could be created that generalizes the values using a mapping of terms. For example, a data set concerning customer transactions may contain a variable *Company* that details the individual customer's company. There may only be a handful of observations for each company. However, this variable could be mapped onto a new variable, *Industries*. The mapping of specific companies onto generalized industries must be defined using a concept mapping (i.e., which company maps onto which industry). Now, when the data set is summarized using the values for the *Industries* variable, meaningful trends may be observed.

3.8 COMBINING VARIABLES

The variable that you are trying to use may not be present in the data set but it may be derived from existing variables. Mathematical operations, such as average or sum, could be applied to one or more variables in order to create an additional variable. For example, a project may be trying to understand issues regarding a particular car's fuel efficiency (*Fuel Efficiency*) using a data set of different journeys in which the fuel level at the start (*Fuel Start*) and end (*Fuel End*) of a trip is measured along with the distance covered (*Distance*). An additional column may be calculated using the following formula:

$$Fuel\ Efficiency = (Fuel\ End - Fuel\ Start)/Distance$$

Different approaches to generating new variables to support model building will be discussed in Chapter 6.

3.9 GENERATING GROUPS

Generally, larger data sets take more computational time to analyze and creating subsets from the data can speed up the analysis. One approach is to take a random subset which is effective where the data set closely matches the target population.

Another reason for creating subsets is when a data set that has been built up over time for operational purposes, but is now to be used to answer an alternative business research question. It may be necessary to select a diverse set of observations that more closely matches the new target population. For example, suppose a car safety organization has been measuring the safety of individual cars based on specific requests from the government. Over time, the government may have requested car safety studies for certain types of vehicles. If the historical data set is to be used to answer questions on the safety of all cars, this data set does not reflect the new target population. However, a subset of the car studies could be selected to represent the more general questions now being asked of the data. The chapter on grouping will discuss how to create diverse data sets when the data does not represent the target population.

A third reason is that when building predictive models from a data set, it is important to keep the models as simple as possible. Breaking the data set down into subsets based on your knowledge of the data may allow you to create several simpler models. For example, a project to model factors that contribute to the price of real estate may use a data set of nationwide house prices and associated factors. However, your knowledge of the real estate market suggests that factors contributing to housing prices are contingent upon the area in which the house is located. Factors that contribute to house prices in coastal locations are different from factors that contribute to house prices in the mountains. It may make sense in this situation to divide the data set up into smaller sets based on location and model these locales separately. When doing this type of subsetting, it is important to note the criteria you are using to subset the data. The specific criteria is needed when data to be predicted are presented for modeling by assigning the data to one or more models. In situations where multiple predictions are generated for the same unknown observation, a method for consolidating these predictions is required.

3.10 PREPARING UNSTRUCTURED DATA

In many disciplines, the focus of a data analysis or data mining project is not a simple data table of observations and variables. For example, in the life sciences, the focus of the analysis is genes, proteins, biological pathways, and chemical structures. In other disciplines, the focus of the analysis could be documents, web logs, device readouts, audio or video information, and so on. In the analysis of these types of data, a preliminary step is often the computational generation of different attributes relevant to the problem. For example, when analyzing a data set of chemicals, an initial step is to generate variables based on the composition of the chemical such as its molecular weight or the presence or absence of molecular components.

TABLE 3.3 Table of Patient Records

Name	Age	Gender	Blood Group	Weight (kg)	Height (m)	Systolic Blood Pressure (mm Hg)	Diastolic Blood Pressure (mm Hg)	Temperature (°F)	Diabetes
P. Lee	35	Female	A Rh+	50	1.52	68	112	98.7	0
R. Jones	52	Male	O Rh−	115	1.77	110	154	98.5	1
J. Smith	45	Male	O Rh+	96	1.83	88	136	98.8	0
A. Patel	70	Female	O Rh−	41	1.55	76	125	98.6	0
M. Owen	24	Male	A Rh−	79	1.82	65	105	98.7	0
S. Green	43	Male	O Rh−	109	1.89	114	159	98.9	1
N. Cook	68	Male	A Rh+	73	1.76	108	136	99.0	0
W. Hands	77	Female	O Rh−	104	1.71	107	145	98.3	1
P. Rice	45	Female	O Rh+	64	1.74	101	132	98.6	0
F. Marsh	28	Male	O Rh+	136	1.78	121	165	98.7	1

EXERCISES

A set of 10 hypothetical patient records from a large database is presented in Table 3.3. Patients with a diabetes value of 1 have type-II diabetes and patients with a diabetes value of 0 do not have type-II diabetes.

1. Create a new column by normalizing the *Weight* (*kg*) variable into the range 0–1 using the min–max normalization.
2. Create a new column by binning the *Weight* (*kg*) variable into three categories: low (less than 60 kg), medium (60–100 kg), and high (greater than 100 kg).
3. Create an aggregated column, *body mass index* (*BMI*), which is defined by the formula:

$$\text{BMI} = \frac{\text{Weight(kg)}}{(\text{Height(m)})^2}$$

FURTHER READING

For additional data preparation approaches including the handling of missing data see Pearson (2005), Pyle (1999), and Dasu & Johnson (2003).

CHAPTER 4

UNDERSTANDING RELATIONSHIPS

4.1 OVERVIEW

A critical step in making sense of data is an understanding of the relationships between different variables. For example, is there a relationship between interest rates and inflation or education level and income? The existence of an association between variables does not imply that one variable causes another. These relationships or associations can be established through an examination of different summary tables and data visualizations as well as calculations that measure the strength and confidence in the relationship. The following sections examine a number of ways to understand relationships between pairs of variables through data visualizations, tables that summarize the data, and specific calculated metrics. Each approach is driven by how the variables have been categorized such as the scale on which they are measured. The use of data visualizations is important as it takes advantage of the human visual system's ability to recognize complex patterns in what is seen graphically.

Making Sense of Data I: A Practical Guide to Exploratory Data Analysis and Data Mining,
Second Edition. Glenn J. Myatt and Wayne P. Johnson.
© 2014 John Wiley & Sons, Inc. Published 2014 by John Wiley & Sons, Inc.

4.2 VISUALIZING RELATIONSHIPS BETWEEN VARIABLES

4.2.1 Scatterplots

Scatterplots can be used to identify whether a relationship exists between two continuous variables measured on the ratio or interval scales. The two variables are plotted on the *x*-and *y*-axis. Each point displayed on the scatterplot is a single observation. The position of the point is determined by the value of the two variables. The scatterplot in Figure 4.1 presents hundreds of observations on a single chart.

Scatterplots allow you to see the type of relationship that may exist between two variables. A *positive relationship* results when higher values in the first variable coincide with higher values in the second variable and lower values in the first variable coincide with lower values in the second variable (the points in the graph are trending upward from left to right). *Negative relationships* result when higher values in the first variable coincide with lower values in the second variable and lower values in the first variable coincide with higher values in the second variable (the points are trending downward from left to right). For example, the scatterplot in Figure 4.2 shows that the relationship between *petal length* (*cm*) and *sepal length* (*cm*) is positive.

The nature of the relationships—linearity or nonlinearity—is also important. A linear relationship exists when a second variable changes *proportionally* in response to changes in the first variable. A *nonlinear* relationship is drawn as a curve indicating that as the first variable changes, the change in the second variable is not proportional. In Figure 4.2 the

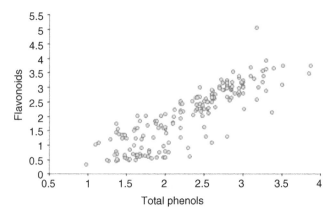

FIGURE 4.1 Example of a scatterplot where each point corresponds to an observation.

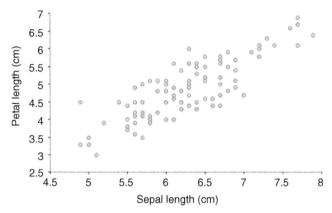

FIGURE 4.2   A scatterplot showing a positive relationship.

relationship is primarily linear—as *sepal length* (*cm*) increases, *petal length* (*cm*) increases proportionally. A scatterplot can also show if there are points (e.g., X on Figure 4.3) that do not follow this linear relationship. These are referred to as outliers.

Scatterplots may also indicate *negative* relationships. For example, it can be seen in Figure 4.4 that as the values for *weight* increase, the values for *MPG* decrease. In situations where the relationship between the variables is more complex, there may be a combination of positive and negative relationships at various points. In Figure 4.4, the points follow a curve indicating that there is also a *nonlinear* relationship between the two variables—as *weight* increases *MPG* decreases, but the rate of decrease is not proportional.

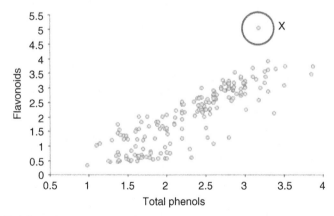

FIGURE 4.3   Observation (marked as X) that does not follow the relationship.

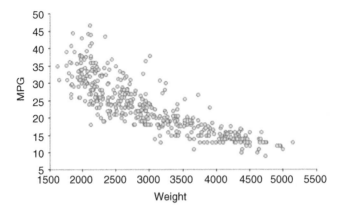

FIGURE 4.4   Scatterplot showing a negative nonlinear relationship.

Scatterplots can also show the lack of any relationship. In Figure 4.5, the points scattered throughout the graph indicates that there is no obvious relationship between *Alcohol* and *Nonflavonoid phenols* in this data set.

4.2.2   Summary Tables and Charts

A simple *summary table* is a common way of understanding the relationship between two variables where at least one of the variables is discrete. For example, a national retail company may have collected information on the sale of individual products for every store. To summarize the performance of these stores, they may wish to generate a summary table to communicate the average sales per store.

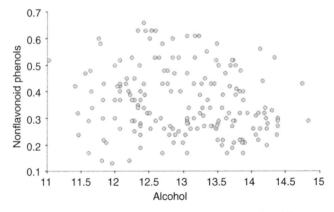

FIGURE 4.5   Scatterplot showing no relationships.

Class	Count	Minimum (petal width (cm))	Maximum (petal width (cm))	Mean (petal width (cm))	Median (petal width (cm))	Standard deviation (petal width (cm))
Iris-setosa	50	0.1	0.6	0.244	0.2	0.107
Iris-versicolor	50	1	1.8	1.33	1.3	0.198
Iris-virginica	50	1.4	2.5	2.03	2	0.275

FIGURE 4.6 Example of a summary table.

A single categorical variable (or a continuous variable converted into categories) is used to group the observations, and each row of the table represents a single group. Summary tables will often show a count of the number of observations (or percentage) that has the particular value (or range). Any number of other columns can be shown alongside to summarize the second variable. Since each row now refers to a set of observations, other columns of variables included in the table must also contain summary information. Descriptive statistics that summarize a set of observations can be used including mean, median, mode, sum, minimum, maximum, variance, and standard deviation.

In Figure 4.6, the relationship between two variables *class* and *petal width (cm)* is shown from Fisher (1936). The *class* variable is a discrete variable (nominal) that can take values "Iris-setosa," "Iris-versicolor," and "Iris-virginica" and each of these values is shown in the first column. There are 50 observations that correspond to each of these values and each row of the table describes the corresponding 50 observations. Each row is populated with summary information about the second variable (*petal width (cm)*) for the set of 50 observations. In this example, the minimum and maximum values are shown alongside the mean, median, and standard deviation. It can be seen that the class "Iris-setosa" is associated with the smallest petal width with a mean of 0.2. The set of 50 observations for the class "Iris-versicolor" has a mean of 1.33, and the class "Iris-virginica" has the highest mean of 2.03.

It can also be helpful to view this information as a graph. For example in Figure 4.7, a bar chart is drawn with the *x*-axis showing the three nominal *class* categories and the *y*-axis showing the mean value for *petal width (cm)*. The relative mean values of the three classes can be easily seen using this view.

More details on the frequency distribution of the three separate sets can be seen by using box plots for each category as shown in Figure 4.8. Again, the *x*-axis represents the three classes; however, the *y*-axis is now the original values for the petal width of each observation. The box plot illustrates for each individual class how the observations are distributed relative to the other class. For example, the 50 observations in the "Iris-setosa" class have no overlapping values with either of the other two

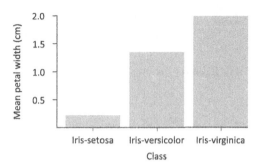

FIGURE 4.7   Example of a summary graph.

classes. In contrast, the frequency distribution of the "Iris-versicolor" and the "Iris-virginica" show an amount of overlap at the extreme values.

Summary tables can also be used to show the relationship between ordinal variables and another variable. In the example shown in Figure 4.9, three ordered categories are used to group the observations. Since the categories can be ordered, it is possible to see how the mean weight changes as the MPG category increases. It is clear from this table that as *MPG* categories increases the mean *weight* decreases.

The same information can be seen as a histogram and a series of box plots. By ordering the categories on the *x*-axis and plotting the information, the trend can be seen more easily, as shown in Figure 4.10.

In many situations, a binary variable is used to represent a variable with two possible values, with 0 representing one value and 1 the other. For example, 0 could represent the case where a patient has a specific infection and 1 the case where the patient does not. To illustrate, a new blood test is being investigated to predict whether a patient has a specific

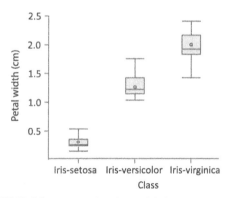

FIGURE 4.8   Example of a multiple box plot graph.

MPG categories	Count	Mean (weight)
0–20 MPG	151	3832
20–30 MPG	151	2612
30–50 MPG	90	2157

FIGURE 4.9 Example of summary table where the categorical variable is ordinal.

type of infection. This test calculates a value between −5.0 and +1.0, with higher values indicating the patient has the infection and lower values indicating the patient does not have the infection. The results of this test are summarized in Figure 4.11, which shows the data categorized by the two patient classes. There are 23 patients in the data that did not have the infection (*infection* = 0) and 32 that did have the infection (*infection* = 1). It can be seen that there is a difference between the blood test results for the two groups. The mean value of the blood test in the group with the infection is −3.05, whereas the mean value of the blood test over patients without the infection is −4.13. Although there is a difference, the overlap of the blood test results between the two groups makes it difficult to interpret the results.

A different way to assess this data would be to use the *blood test* results to group the summary table and then present descriptive statistics for the *infection* variable. Since the blood test values are continuous, the first step is to organize the blood test results into ranges. Range boundaries were set at −5, −4, −3, −2, −1, 0, +1, and +2 and groups generated as shown in Figure 4.12. The mean of the binary variable *infection* is shown in the

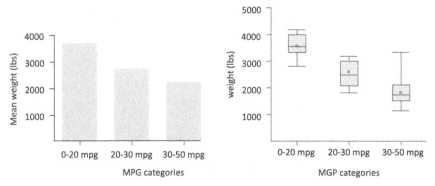

FIGURE 4.10 Graph of summary data for an ordinal variable against a continuous (weight).

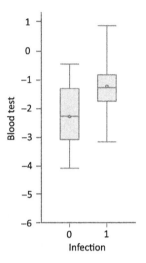

Infection	Count	Minimum (blood test)	Maximum (blood test)	Mean (blood test)	Standard deviation (blood test)
0	23	−4.13	−0.393	−2.18	1.04
1	32	−3.05	0.893	−1.09	0.834

FIGURE 4.11   Summary table and corresponding box plot chart to summarizing the results of a trial for a new blood test to predict an infection.

summary table corresponding to each of the groups and also plotted using a histogram. If all observations in a group were 0, then the mean would be 0; and if all observations in a group were 1, then the mean would be 1. Hence the upper and lower limits on these mean values are 0 and 1. This table and chart more clearly shows the relationship between these

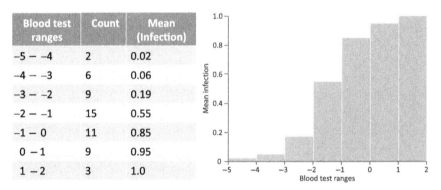

Blood test ranges	Count	Mean (Infection)
−5 − −4	2	0.02
−4 − −3	6	0.06
−3 − −2	9	0.19
−2 − −1	15	0.55
−1 − 0	11	0.85
0 − 1	9	0.95
1 − 2	3	1.0

FIGURE 4.12   Summary table and histogram using continuous data that has been binned to generate the group summarized using the binary variable.

Infection Class

Test Results		Infection negative	Infection positive	Totals
	Blood test negative	17	10	27
	Blood test positive	6	22	28
	Totals	23	32	55

FIGURE 4.13   Contingency table showing the relationship between two dichotomous variables.

two variables. For lower values of the blood test (from −5 to −2) there are few cases of the infection—the test correctly predicts the majority of these cases as negative. Similarly, for the range −1 to 2 most of these cases are positive and the test would correctly predict most of these cases as positive. However, for the range of blood test from −2 to −1, 55% are predicted positive and 45% are predicted negative (based on the mean value for this range). For this range of values the blood test performs poorly. It would be reasonable to decide that for a test result in this range, the results should not be trusted.

4.2.3 Cross-Classification Tables

Cross-classification tables or *contingency tables* provide insight into the relationship between two categorical variables (or non-categorical variables transformed to categorical variables). A variable is often dichotomous; however, a contingency table can represent variables with more than two values. Figure 4.13 provides an example of a contingency table for two variables over a series of patients: *Test Results* and *Infection Class*. The variable *Infection Class* identifies whether a patient has the specific infection and can take two possible values ("Infection negative" and "Infection positive"). The corresponding *Test Results* identified whether the blood test results were positive ("Blood test positive") or negative ("Blood test negative"). In this data set there were 55 observations (shown in the bottom-right cell). Totals for the *Test Results* values are shown in the rightmost column labeled "Totals" and totals for *Infection Class* values are shown on the bottom row also labeled "Totals." In this example, there are 27 patients where the values for Test Results are "Blood test negative" and 28 that are "Blood test positive" (shown in the right column). Similarly, there are 23 patients that are categorized as "Infection negative" and 32

Gender

Age-group	Male	Female	Totals
10–19	847	810	1657
20–29	4878	3176	8054
30–39	6037	2576	8613
40–49	5014	2161	7175
50–59	3191	1227	4418
60–69	1403	612	2015
70–79	337	171	508
80–89	54	24	78
90–99	29	14	43
Total	21,790	10,771	32,561

FIGURE 4.14 Contingency table illustrating the number of females and males in each age-group.

as "Infection positive." The table cells in the center of the table show the number of patients that correspond to pairs of values. For example, there are 17 patients who had a negative blood test result and did not have the infection ("Infection negative"), 10 patients who had a negative blood test result but had an infection, 6 who had a positive blood test result and did not have an infection, and 22 who had a positive blood test and had an infection. Contingency tables provide a view of the relationship between two categorical variables. It can be seen from this example that although the new blood test did not perfectly identify the presence or absence of the infection, it correctly classified the presence of the infection for 39 patients (22 + 17) and incorrectly classified the infection in 16 patients (10 + 6).

Contingency tables can be used to understand relationships between categorical (both nominal and ordinal) variables where there are more than two possible values. In Figure 4.14, the data set is summarized using two variables: *gender* and *age-group*. The variable *gender* is dichotomous and the two values ("female" and "male") are shown in the table's header on the *x*-axis. The other selected variable is *age-group*, which has been broken down into nine categories: 10–19, 20–29, 30–39, etc. For each level of each variable, a total is displayed. For example, there are 21,790 observations where *gender* is "male" and there are 1657 observations where age is between 10 and 19. The total number of observations summarized in the table is shown in the bottom right-hand corner (32,561).

4.3 CALCULATING METRICS ABOUT RELATIONSHIPS

4.3.1 Overview

There are many ways to measure the strength of the relationship between two variables. These metrics are usually based on the types of variables being considered, such as a comparison between categorical variables and continuous variables. The following section describes common methods for quantifying the strength of relationships between variables.

4.3.2 Correlation Coefficients

For pairs of variables measured on an interval or ratio scale, a *correlation coefficient* (*r*) can be calculated. This value quantifies the *linear relationship* between the variables by generating values from −1.0 to +1.0. If the optimal straight line is drawn through the points on a scatterplot, the value of *r* reflects how closely the points lie to this line. Positive numbers for *r* indicate a positive correlation between the pair of variables, and negative numbers indicate a negative correlation. A value of *r* close to 0 indicates little or no relationship between the variables.

For example, the two scatterplots shown in Figure 4.15 illustrate different values for *r*. The first graph illustrates a strong positive correlation because the points lie relatively close to an imaginary line sloping upward from left to right through the center of the points; the second graph illustrates a weaker correlation.

The formula used to calculate *r* is shown here:

$$r = \frac{\sum_{i=1}^{n}(x_i - \bar{x})(y_i - \bar{y})}{(n-1)s_x s_y}$$

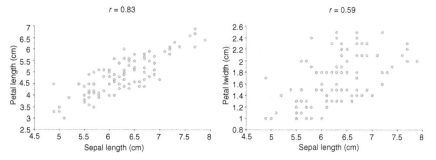

FIGURE 4.15 Scatterplots illustrate values for the correlation coefficient (*r*).

TABLE 4.1 Table of Data with Values for the x and y Variables

x	y
92	6.3
145	7.8
30	3.0
70	5.5
75	6.5
105	5.5
110	6.5
108	8.0
45	4.0
50	5.0
160	7.5
155	9.0
180	8.6
190	10.0
63	4.2
85	4.9
130	6
132	7

where x and y are variables, x_i are the individual values of x, y_i are the individual values of y, $\bar{x}$ is the mean of the x variable, $\bar{y}$ is the mean of the y variable, s_x and s_y are the standard deviations of the variables x and y, respectively, and n is the number of observations. To illustrate the calculation, two variables (x and y) are used and shown in Table 4.1. The scatterplot of the two variables indicates a positive correlation between them, as shown in Figure 4.16. The specific value of r is calculated using Table 4.2:

$$r = \frac{\sum_{i=1}^{n}(x_i - \bar{x})(y_i - \bar{y})}{(n-1)s_x s_y}$$

$$r = \frac{1357.06}{(18-1)(47.28)(1.86)}$$

$$r = 0.91$$

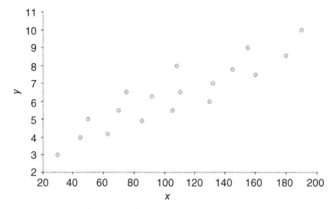

FIGURE 4.16 Scatterplot to illustrate the relationship between the x and y variables.

TABLE 4.2   Table Showing the Calculation of the Correlation Coefficient

x_i	y_i	$(x_i - \bar{x})$	$(y_i - \bar{y})$	$(x_i - \bar{x})(y_i - \bar{y})$
92	6.3	−14.94	−0.11	1.64
145	7.8	38.06	1.39	52.90
30	3	−76.94	−3.41	262.37
70	5.5	−36.94	−0.91	33.62
75	6.5	−31.94	0.09	−2.87
105	5.5	−1.94	−0.91	1.77
110	6.5	3.06	0.09	0.28
108	8	1.06	1.59	1.69
45	4	−61.94	−2.41	149.28
50	5	−56.94	−1.41	80.04
160	7.5	53.06	1.09	58.07
155	9	48.06	2.59	124.68
180	8.6	73.06	2.19	160.00
190	10	83.06	3.59	298.19
63	4.2	−43.94	−2.21	97.11
85	4.9	−21.94	−1.51	33.13
130	6	23.06	−0.41	−9.45
132	7	25.06	0.59	14.79

$\bar{x} = 106.94$ $\bar{y} = 6.41$ $Sum = 1,357.06$
$s_x = 47.28$ $s_y = 1.86$

4.3.3 Kendall Tau

Kendall Tau is another approach for measuring associations between pairs of variables. It is based on a ranking of the observations for two variables. This ranking can be derived by ordering the values and then replacing the actual values with a rank from 1 to n (where n is the number of observations in the data set). The overall formula is based on counts of *concordant* and *discordant* pairs of observations. Table 4.3 is used to illustrate Kendall Tau using two variables, *Variable X* and *Variable Y*, containing a ranking with 10 observations (A through J).

A pair of observations is *concordant* if the difference of one of the two variable's values is in the same direction as the difference of the other variable's values. For example, to determine if the observations A and B are concordant, we compare the difference of the values for *Variable X* $(X_B - X_A$ or $2 - 1 = 1)$ with the difference of the values for *Variable Y* $(Y_B - Y_A$ or $4 - 2 = 2)$. Since these differences are in the same direction— 1 and 2 are both positive—the observations A and B are concordant. We would get the same result if we compared the difference of $(X_A - X_B)$ and $(Y_A - Y_B)$. In this case the differences would both be negative, but still in the same direction. Calculated either way, the observations A and B are concordant. A *discordant* pair occurs when the differences of the two variables' values move in opposite directions. The pair of observations B and C illustrates this. The difference of the values for *Variable X* $(X_C - X_B$ or $3 - 2 = 1)$ compared with the difference of the values for *Variable Y* $(Y_C - Y_B$ or $1 - 4 = -3)$ are in different directions: the first is positive and the second negative. The pair of observations B and C is discordant.

TABLE 4.3 Data Table of Rankings for Two Variables

Observation name	Variable X	Variable Y
A	1	2
B	2	4
C	3	1
D	4	3
E	5	6
F	6	5
G	7	7
H	8	8
I	9	10
J	10	9

TABLE 4.4   Calculation of Concordant and Discordant Pairs

Observation Name	Variable X	Variable Y	Concordant Pairs	Discordant Pairs
A	1	2	8	1
B	2	4	6	2
C	3	1	7	0
D	4	3	6	0
E	5	6	4	1
F	6	5	4	0
G	7	7	2	1
H	8	8	2	0
I	9	10	0	1
J	10	9	0	0
		Sum:	39	6

In Table 4.4, the observations A–J are ordered using *Variable X*, and each unique pair of observations is compared. For example, observation A is compared with all other observations (B, C, ... , J) and the number of concordant and discordant pairs are summed as shown in the last two columns of Table 4.4. For observations A, there are eight concordant pairs (A–B, A–D, A–E, A–F, A–G, A–H, A–I, A–J) and one discordant pair (A–C). To make sure that each pair of observations is considered in this process, this is repeated for all other observations. Observation B is compared to observations C through J, observation C compared to D though J, and so on. The number of concordant and discordant pairs is shown for each observation, and the sum of the concordant and discordant pairs is also computed.

A final *Kendall Tau* measure is computed based on these computed sums. Kendall Tau measures associations between variables with 1 indicating a perfect ranking and −1 a perfect disagreement of the rankings. A zero value—assigned when the ranks are tied—indicates a lack of association, or in other words, that the two variables are independent. The simplest form of the Kendall Tau calculation is referred to as Tau A and has the formula:

$$\tau_A = \frac{n_c - n_d}{n(n-1)/2}$$

where n_c and n_d are the number of concordant and discordant pairs, respectively, and n is the number of observations. In this example, the Tau

A (τ_A) would be:

$$\tau_A = \frac{39 - 6}{45}$$

$$\tau_A = 0.73$$

In most practical situations, there are ties based on either variable. In these situations, the formula Tau B is often used. It considers the ties in the first variable (t_x) and the ties in the second variable (t_y) and is computed using the following formulas:

$$\tau_B = \frac{n_c - n_d}{\sqrt{(n_c + n_d + t_x)(n_c + n_d + t_y)}}$$

In most software applications the Kendall Tau B function is used.

4.3.4 *t*-Tests Comparing Two Groups

In Chapter 2, we used a hypothesis test (*t*-test) to determine whether the mean of a variable was sufficiently different from a specified value as to be considered statistically significant, meaning that it would be unlikely that the difference between the value and the mean was due to normal variation or chance. This concept can be extended to compare the mean values of two subsets. We can explore if the means of two groups are different enough to say the difference is significant, or conclude that a difference is simply due to chance. This concept is similar to the description in Chapter 2; however, different formulas are utilized.

In looking at the difference between two groups, we need to not only take into account the values for the mean values of the two groups, but also the deviation of the data for the two groups.

The formula is used where it is assumed that the value being assessed across the two groups is both independently and normally distributed and variances between the two groups are either equal or similar. The formula takes into account the difference between the two groups as well as information concerning the distribution of the two groups:

$$T = \frac{\bar{x}_1 - \bar{x}_2}{s_p\sqrt{\dfrac{1}{n_1} + \dfrac{1}{n_2}}}$$

where $\bar{x}_1$ is the mean value of the first group, $\bar{x}_2$ is the mean value of the second group, and n_1 and n_2 are the number of observations in the first and second group respectively, and s_p is an estimate of the standard deviation (pooled estimate). The s_p is calculated using the following formula:

$$s_p = \sqrt{\frac{(n_1 - 1)s_1^2 + (n_2 - 1)s_2^2}{n_1 + n_2 - 2}}$$

where n_1 and n_2 are again the number of observations in group 1 and group 2 respectively, and s_1^2, s_2^2 are the calculated variances for group 1 and group 2. This formula follows a t-distribution, with the number of degrees of freedom (df) calculated as

$$df = n_1 + n_2 - 2$$

where it cannot be assumed that the variances across the two groups are equal, another formula is used:

$$T = \frac{\bar{x}_1 - \bar{x}_2}{\sqrt{\frac{s_1^2}{n_1} + \frac{s_2^2}{n_1}}}$$

where $\bar{x}_1$ and $\bar{x}_2$ are the average values for the two groups (1 and 2), s_1^2 and s_2^2 are the calculated variances for the two groups, and n_1 and n_2 are the number of observations in the two groups.

Again, it follows a t-distribution and the number of degrees of freedom (df) is calculated using the following formula:

$$df = \frac{\left(\frac{s_1^2}{n_1} + \frac{s_1^2}{n_2}\right)^2}{\frac{1}{n_1 - 1}\left(\frac{s_1^2}{n_1}\right)^2 + \frac{1}{n_2 - 1}\left(\frac{s_2^2}{n_2}\right)^2}$$

These t-values will be positive if the mean of group 1 is larger than the mean of group 2 and negative if the mean of group 2 is larger than the mean of group 1. In a similar manner as described in Chapter 2, these t-values can be used in a hypothesis test where the null hypothesis states that the two means are equal and the alternative hypothesis states that the two means are not equal. This t-value can be used to accept or reject the null hypothesis as well as calculate a p-value, either using a computer or a statistical table, in a manner similar to that described in Chapter 2.

TABLE 4.5   Calls Processed by Different Call Centers

Call Center A	Call Center B	Call Center C	Call Center D
136	124	142	149
145	131	145	157
139	128	139	154
132	130	145	155
141	129	143	151
143	135	141	156
138	132	138	
139		146	

4.3.5 ANOVA

The following section reviews a technique called the completely randomized *one-way analysis of variance* that compares the means of three or more different groups. The test determines whether there is a difference between the groups. This method can be applied to cases where the groups are independent and random, the distributions are normal and the populations have similar variances. For example, an online computer retail company has call centers in four different locations. These call centers are approximately the same size and handle a certain number of calls each day. An analysis of the different call centers based on the average number of calls processed each day is required to understand whether one or more of the call centers are under- or over-performing. Table 4.5 illustrates the calls serviced daily.

As with other hypothesis tests, it is necessary to state a null and alternative hypothesis. Generally, the hypothesis statement will take the standard form:

H_0: The means are equal.
H_a: The means are not equal.

To determine whether a difference exists between the means or whether the difference is due to random variation, we must perform a hypothesis test. This test will look at both the variation *within the groups* and the variation *between the groups*. The test performs the following steps:

1. Calculate group means and variance.
2. Determine the within-group variation.

TABLE 4.6   Calculating Means and Variances

	Call Center A	Call Center B	Call Center C	Call Center D	Groups ($k = 4$)
	136	124	142	149	
	145	131	145	157	
	139	128	139	154	
	132	130	145	155	
	141	129	143	151	
	143	135	141	156	
	138	132	138		
	139		146		
Count	8	7	8	6	*Total count N = 29*
Mean	139.1	129.9	142.4	153.7	
Variance	16.4	11.8	8.6	9.5	

3. Determine the between-group variation.
4. Determine the F-statistic, which is based on the between-group and within group ratio.
5. Test the significance of the F-statistic.

The following sections describe these steps in detail:

Calculate group means and variances
In Table 4.6, for each call center a count along with the mean and variance has been calculated. In addition, the total number of groups ($k = 4$) and the total number of observations ($N = 29$) is listed. An average of all values ($\bar{\bar{x}} = 140.8$) is calculated by summing all values and dividing it by the number of observations:

$$\bar{\bar{x}} = \frac{136 + 145 + \ldots + 151 + 156}{29} = 140.8$$

Determine the within-group variation
The variation within groups is defined as the within-group variance or *mean square within* (*MSW*). To calculate this value, we use a weighted sum of the variance for the individual groups. The weights are based on the number of observations in each group. This sum is divided by the

number of degrees of freedom calculated by subtracting the number of groups (k) from the total number of observations (N):

$$MSW = \frac{\sum_{i=1}^{k}(n_i - 1)s_i^2}{N - k}$$

In this example:

$$MSW = \frac{(8-1) \times 16.4 + (7-1) \times 11.8 + (8-1) \times 8.6 + (6-1) \times 9.5}{(29-4)}$$

$$MSW = 11.73$$

Determine the between-group variation

Next, the between-group variation or *mean square between* (*MSB*) is calculated. The *MSB* is the variance between the group means. It is calculated using a weighted sum of the squared difference between the group mean ($\bar{x}_i$) and the average of all observations ($\bar{\bar{x}}$). This sum is divided by the number of degrees of freedom. This is calculated by subtracting one from the number of groups (k). The following formula is used to calculate the *MSB*:

$$MSB = \frac{\sum_{i=1}^{k}n_i(\bar{x}_i - \bar{\bar{x}})^2}{k - 1}$$

where n_i is the number for each group and $\bar{x}_i$ is the average for each group.

In this example,

$$MSB = \frac{\begin{array}{c}(8 \times (139.1 - 140.8)^2) + (7 \times (129.9 - 140.8)^2)\\ + (8 \times (142.4 - 140.8)^2) + (6 \times (153.7 - 140.8)^2)\end{array}}{4 - 1}$$

$$MSB = 624.58$$

Determine the F-statistic

The *F*-statistic is the ratio of the MSB and the MSW:

$$F = \frac{MSB}{MSW}$$

In this example:

$$F = \frac{624.58}{11.73}$$

$$F = 53.25$$

Test the significance of the F-statistic

Before we can test the significance of this value, we must determine the degrees of freedom (df) for the two mean squares (within and between). The degrees of freedom for the MSW (df_{within}) is calculated using the following formula:

$$df_{within} = N - k$$

where N is the total number of observations in all groups and k is the number of groups.

The degrees of freedom for the MSB ($df_{between}$) is calculated using the following formula:

$$df_{between} = k - 1$$

where k is the number of groups.

In this example,

$$df_{between} = 4 - 1 = 3$$

$$df_{within} = 29 - 4 = 25$$

We already calculated the F-statistic to be 53.39. This number indicates that the mean variation between groups is much greater than the mean variation within groups due to errors. To test this, we look up the critical F-statistic from an F-table (see the Further Readings section). To find this critical value we need α (confidence level), v_1 ($df_{between}$), and v_2 (df_{within}). The critical value for the F-statistic is 3.01 (when α is 0.05). Since the calculated F-statistic is greater than the critical value, we reject the null hypothesis. The means for the different call centers are not equal.

4.3.6 Chi-Square

The chi-square test for indepedence is a hypothesis test for use with variables measured on a nominal or ordinal scale. It allows an analysis of whether there is a relationship between two categorical variables. As with

TABLE 4.7   Calculation of Chi-Square

k	Category	Observed (O)	Expected (E)	$(O - E)^2/E$
1	$r =$ Brand X, $c = 43221$	5,521	4,923	72.6
2	$r =$ Brand Y, $c = 43221$	4,597	4,913	20.3
3	$r =$ Brand Z, $c = 43221$	4,642	4,925	16.3
4	$r =$ Brand X, $c = 43026$	4,522	4,764	12.3
5	$r =$ Brand Y, $c = 43026$	4,716	4,754	0.3
6	$r =$ Brand Z, $c = 43026$	5,047	4,766	16.6
7	$r =$ Brand X, $c = 43212$	4,424	4,780	26.5
8	$r =$ Brand Y, $c = 43212$	5,124	4,770	26.3
9	$r =$ Brand Z, $c = 43212$	4,784	4,782	0.0008
				Sum = 191.2

other hypothesis tests, it is necessary to state a null and alternative hypothesis. Generally, these hypothesis statements are as follows:

H_0: There is no relationship.
H_a: There is a relationship.

Using Table 4.7, we will look at whether a relationship exists between where a consumer lives (represented by a zip code) and the brand of washing powder they buy (brand X, brand Y, and brand Z). The "r" and "c" refer to the row (r) and column (c) in a contingency table.

The Chi-Square test compares the observed frequencies with the expected frequencies. The expected frequencies are calculated using the following formula:

$$E_{r,c} = \frac{r \times c}{n}$$

where $E_{r,c}$ is the expected frequency for a particular cell in a contingency table, r is the row count, c is the column count and n is the total number of observations in the sample.

For example, to calculate the expected frequency for the table cell where the washing powder is brand X and the zip code is 43221 would be

$$E_{\text{Brand\_X},43221} = \frac{14,467 \times 14,760}{43,377}$$

$$E_{\text{Brand\_X},43221} = 4,923$$

The Chi-Square test (χ^2) is computed with the following equation:

$$\chi^2 = \sum_{i=1}^{k} \frac{(O_i - E_i)^2}{E_i}$$

where k is the number of all categories, O_i is the observed cell frequency and E_i is the expected cell frequency. The test is usually performed when all observed cell frequencies are greater than 10. Table 4.7 shows the computed χ^2 for this example.

There is a critical value at which the null hypothesis is rejected (χ_c^2) and this value is found using a standard Chi-Square table (see Further Reading Section). The value is dependent on the degrees of freedom (df), which is calculated:

$$df = (r - 1) \times (c - 1)$$

For example, the number of degrees of freedom for the above example is $(3 - 1) \times (3 - 1)$ which equals 4. Looking up the critical value for $df = 4$ and $\alpha = 0.05$, the critical value is 9.488. Since 9.488 is less than the calculated chi-square value of 191.2, we reject the null hypothesis and state that there is a relationship between zip codes and brands of washing powder.

EXERCISES

Table 4.8 shows a series of retail transactions monitored by the main office of a computer store.

1. Generate a contingency table summarizing the variables *Store* and *Product* category.
2. Generate the following summary tables:
   a. Grouping by *Customer* with a count of the number of observations and the sum of *Sale price* ($) for each row.
   b. Grouping by *Store* with a count of the number of observations and the mean *Sale price* ($) for each row.
   c. Grouping by *Product* category with a count of the number of observations and the sum of the *Profit* ($) for each row.
3. Create a scatterplot showing *Sales price* ($) against *Profit* ($).

TABLE 4.8 Retail Transaction Data Set

Customer	Store	Product Category	Product Description	Sale Price ($)	Profit ($)
B. March	New York, NY	Laptop	DR2984	950	190
B. March	New York, NY	Printer	FW288	350	105
B. March	New York, NY	Scanner	BW9338	400	100
J. Bain	New York, NY	Scanner	BW9443	500	125
T. Goss	Washington, DC	Printer	FW199	200	60
T. Goss	Washington, DC	Scanner	BW39339	550	140
L. Nye	New York, NY	Desktop	LR21	600	60
L. Nye	New York, NY	Printer	FW299	300	90
S. Cann	Washington, DC	Desktop	LR21	600	60
E. Sims	Washington, DC	Laptop	DR2983	700	140
P. Judd	New York, NY	Desktop	LR22	700	70
P. Judd	New York, NY	Scanner	FJ3999	200	50
G. Hinton	Washington, DC	Laptop	DR2983	700	140
G. Hinton	Washington, DC	Desktop	LR21	600	60
G. Hinton	Washington, DC	Printer	FW288	350	105
G. Hinton	Washington, DC	Scanner	BW9443	500	125
H. Fu	New York, NY	Desktop	ZX88	450	45
H. Taylor	New York, NY	Scanner	BW9338	400	100

FURTHER READING

For more information on inferential statistics used to assess relationships see Urdan (2010), Anderson et al. (2010), Witte & Witte (2009), and Vickers (2010).

CHAPTER 5

IDENTIFYING AND UNDERSTANDING GROUPS

5.1 OVERVIEW

It is often useful to decompose a data set into simpler subsets to help make sense of the entire collection of observations. These groups may reflect the types of observations found in a data set. For example, the groups might summarize the different types of customers who visit a particular shop based on collected demographic information. Finding subgroups may help to uncover relationships in the data such as groups of consumers who buy certain combinations of products. The process of grouping a data set may also help identify rules from the data, which can in turn be used to support future decisions. For example, the process of grouping historical data can be used to understand which combinations of clinical treatments lead to the best patient outcomes. These rules can then be used to select an optimal treatment plan for new patients with the same symptoms. Finally, the process of grouping also helps discover observations dissimilar from those in the major identified groups. These outliers should be more closely examined as possible errors or anomalies.

Making Sense of Data I: A Practical Guide to Exploratory Data Analysis and Data Mining,
Second Edition. Glenn J. Myatt and Wayne P. Johnson.
© 2014 John Wiley & Sons, Inc. Published 2014 by John Wiley & Sons, Inc.

Class	Count	Mean (petal width (cm))
Iris-setosa	50	0.244
Iris-versicolor	50	1.33
Iris-virginica	50	2.03

FIGURE 5.1 Simple summary table showing how the mean petal width changes for the different classes of flowers.

The identification of interesting groups is not only a common deliverable for a data analysis project, but can also support other data mining tasks such as the development of a model to use in forecasting future events (as described in Chapter 6). This is because the process of grouping and interpreting the groups of observations helps the analyst to thoroughly understand the data set which, in turn, supports the model building process. This grouping process may also help to identify specific subsets that lead to simpler and more accurate models than those built from the entire set. For example, in developing models for predicting house prices, there may be groups of houses (perhaps based on locations) where a simple and clear relationship exists between a specific variable collected and the house prices which allows for the construction of specific models for these subpopulations.

The analysis we have described in Chapter 4 looks at the simple relationship between predefined groups of observations—those encoded using a single predefined variable—and one other variable. For example, in looking at a simple categorization such as types of flowers ("Iris-setosa," "Iris-versicolor," and "Iris-virginica"), a question we might ask is how a single variable such as *petal width* varies among different species as illustrated in Figure 5.1.

This can be easily extended to help understand the relationships between groups and multiple variables, as illustrated in Figure 5.2 where three predefined categories are used to group the observations. Summary information on multiple variables is presented (using the mean value in this

Class	Count	Mean (sepal length (cm))	Mean (sepal width (cm))	Mean (petal length (cm))	Mean (petal width (cm))
Iris-setosa	50	5.01	3.42	1.46	0.244
Iris-versicolor	50	5.94	2.77	4.26	1.33
Iris-virginica	50	6.59	2.97	5.55	2.03

FIGURE 5.2 The use of a summary table to understand multiple variables for a series of groups.

example). As described earlier, these tables can use summary statistics (e.g., mean, mode, median, and so on) in addition to graphs such as box plots that illustrate the subpopulations. The tables may also include other metrics that summarize associations in the data, as described in Chapter 4. A variety of graphs (e.g., histograms, box plots, and so on) for each group can also be shown in a table or grid format known as *small multiples* that allows comparison. For example, in Figure 5.3, a series of variables are plotted (*cylinder, displacement, horsepower, acceleration*, and *MPG*) for three groups ("American cars," "European cars," and "Asian cars"), which clearly illustrates changes in the frequency distribution for each of these classes of cars.

Through an interactive technique known as *brushing*, a subset of observations can also be highlighted within a frequency distribution of the whole data set as illustrated using the automobile example in Figure 5.4. The shaded areas of the individual plots are observations where the car's origin is "American." The chart helps visualize how this group of selected cars is associated with lower fuel efficiency. As shown in the graph in the top left plot (*MPG*), the distribution of the group (dark gray) overlays the distribution of all the cars in the data set (light gray).

In this chapter, we will explore different ways to visualize and group observations by looking at multiple variables simultaneously. One approach is based on similarities of the overall set of variables of interest, as in the case of *clustering*. For example, observations that have high values for certain variables may form groups different from those that have low values for the same variables. In this approach, the pattern of values of variables for observations within a group will be similar, even though the individual data values may differ. A second approach is to identify groups based on interesting combinations of predefined categories, as in the case of *association rules*. This more directed approach identifies associations or rules about groups that can be used to support decision making. For example, a rule might be that a group of customers who historically purchased products A, B, and C also purchased product X. A third directed approach, referred to as *decision trees*, groups observations based on a combination of ranges of continuous variables or of specific categories. As an example, a data set of patients could be used to generate a classification of cholesterol levels based on information such as age, genetic predisposition, lifestyle choices, and so on. This chapter describes how each of these approaches calculates groups and explains techniques for optimizing the results. It also discusses the strengths and weaknesses of each approach and provides worked examples to illustrate them.

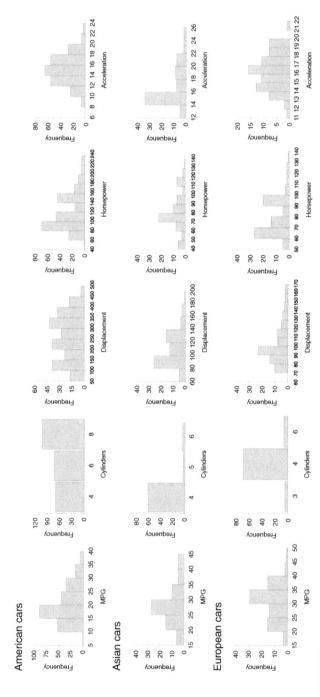

FIGURE 5.3 Matrix showing the frequency distribution for a common set of variables for three groups of cars—American, European, and Asian.

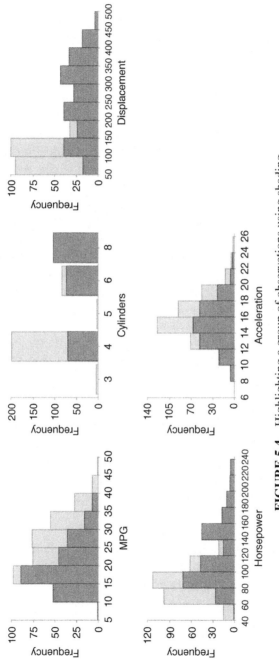

FIGURE 5.4 Highlighting a group of observations using shading.

5.2   CLUSTERING

5.2.1   Overview

For a given data set it is not necessarily known beforehand what groups of observations the entire data set is composed of. For example, in examining customer data collected from a particular store, it is possible to identify and summarize classes of customers directly from the data to answer questions such as "What types of customers visit the store?" Clustering groups data into sets of related observations or clusters, so that observations within each group are more similar to other observations within the group than to observations within other groups. Here, we use the concept of *similarity* abstractly but define it more precisely in Section 5.2.2.

Clustering is an *unsupervised* method for grouping. By unsupervised, we mean that the groups are not known in advance and a goal—a specific variable—is not used to direct how the grouping is generated. Instead, all variables are considered in the analysis. The clustering method chosen to subdivide the data into groups applies an automated procedure to discover the groups based on some criteria and its solution is extracted from patterns or structure existing in the data. There are many clustering methods, and it is important to know that each will group the data differently based on the criteria it uses, regardless of whether meaningful groups exist or not. For clustering, there is no way to measure accuracy and the solution is judged by its "usefulness." For that reason, clustering is used as an open-ended way to explore, understand, and formulate questions about the data in exploratory data analysis.

To illustrate the process of clustering, a set of observations are shown on the scatterplot in Figure 5.5. These observations are plotted using two hypothetical dimensions and the similarity between the observations is proportional to the physical distance between the observations. There are two clear regions that can be considered as clusters: Cluster A and Cluster B, since many of the observations are contained within these two regions on the scatterplot.

Clustering is a flexible approach for grouping. For example, based on the criteria for clustering the observations, observation X was not determined to be a member of cluster A. However, if a more relaxed criterion was used, X may have been included in cluster A. Clustering not only assists in identifying groups of related observations, it can also locate outliers—observations that are not similar to others—since they fall into groups of their own. In Figure 5.5, there are six observations that do not fall within cluster A or B.

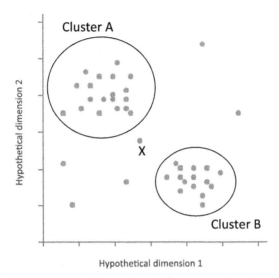

FIGURE 5.5   Illustration of clusters and outliers.

There are two major categories of clustering. Some clustering methods organize data sets hierarchically, which may provide additional insight into the problem under investigation. For example, when clustering genomics data sets, hierarchical clustering may provide insight into the biological functional processes associated with collections of related genes. Other clustering methods partition the data into lists of clusters based on a pre-defined number of groups. For these methods, the speed of computation outweighs the challenge of determining in advance the number of groups that should be used.

There are other factors to consider in choosing and fine-tuning the clustering of a data set. Adjusting the criteria clustering methods use includes options for calculating the similarity between observations and for selecting cluster size. Different problems require different clustering options and may require repeated examination of the results as the options are adjusted to make sense of a particular cluster. Finally, it is important to know the limits of the algorithms. Some clustering methods are time-consuming and, especially for large data sets, may be too computationally expensive to consider, while other methods may have limitations on the number of observations they can process.

To illustrate how clustering works, two clustering techniques will be described in this section: *hierarchical agglomerative clustering* and *k-means clustering*. References to additional clustering methods will be provided in the Further Reading section of this chapter. All approaches to

TABLE 5.1   Table Showing Two Observations A and B

Observation ID	Variable 1	Variable 2
A	2	3
B	7	8

clustering require a formal approach to defining how similar two observations are to each other as measured by the *distance* between them, and this is described in Section 5.2.2.

5.2.2 Distances

A method of clustering needs a way to measure how *similar* observations are to each other. To calculate similarity, we need to compute the *distance* between observations. To illustrate the concept of distance we will use a simple example with two observations and two variables (see Table 5.1). We can see the physical distance between the two observations by plotting them on the following scatterplot (Figure 5.6). In this example, the distance between the two observations is calculated using simple trigonometry:

$$x = 7 - 2 = 5$$
$$y = 8 - 3 = 5$$
$$d = \sqrt{x^2 + y^2} = \sqrt{25 + 25} = 7.07$$

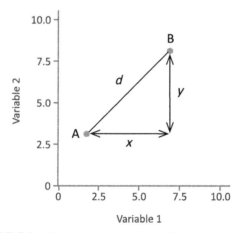

FIGURE 5.6   Distance between two observations A and B.

TABLE 5.2 Three Observations with Values for Five Variables

ID	Variable 1	Variable 2	Variable 3	Variable 4	Variable 5
A	0.7	0.8	0.4	0.5	0.2
B	0.6	0.8	0.5	0.4	0.2
C	0.8	0.9	0.7	0.8	0.9

It is possible to calculate distances between observations with more than two variables by extending this approach. This calculation is called the *Euclidean* distance (d) and its general formula is

$$d = \sqrt{\sum_{i=1}^{n} (p_i - q_i)^2}$$

This formula calculates the distance between two observations p and q where each observation has n variables. To illustrate the Euclidean distance calculation for observations with more than two variables, we will use Table 5.2.

The Euclidean distance between A and B is

$$d_{A-B} = \sqrt{(0.7 - 0.6)^2 + (0.8 - 0.8)^2 + (0.4 - 0.5)^2 + (0.5 - 0.4)^2 + (0.2 - 0.2)^2}$$
$$d_{A-B} = 0.17$$

The Euclidean distances between A and C is

$$d_{A-C} = \sqrt{(0.7 - 0.8)^2 + (0.8 - 0.9)^2 + (0.4 - 0.7)^2 + (0.5 - 0.8)^2 + (0.2 - 0.9)^2}$$
$$d_{A-C} = 0.83$$

The Euclidean distance between B and C is

$$d_{B-C} = \sqrt{(0.6 - 0.8)^2 + (0.8 - 0.9)^2 + (0.5 - 0.7)^2 + (0.4 - 0.8)^2 + (0.2 - 0.9)^2}$$
$$d_{B-C} = 0.86$$

The distance between A and B is 0.17, whereas the distance between A and C is 0.83, which indicates that there is more similarity between observations A and B than A and C. C is not closely related to either A or B. This can be seen in Figure 5.7 where the values for each variable are plotted along the horizontal axis and the height of the bar measured against the vertical axis represents the data value. The shape of histograms A and B are similar, whereas the shape of histogram C is not similar to A or B.

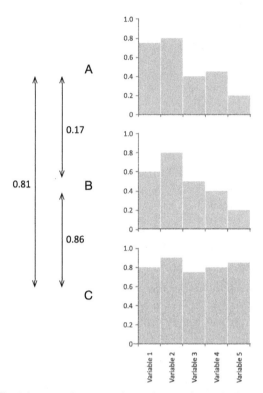

FIGURE 5.7 Distances between three observations: A–B, B–C, and A–C.

Sometimes a distance metric can only be used for a particular type of variable. The Euclidean distance metric can be used only for numerical variables. Other metrics are needed for binary variables and one of which is the *Jaccard* distance. This approach is based on the number of common or different 0/1 values between corresponding variables across each pair of observations using the following counts:

- $Count_{11}$: Count of all variables that are 1 in "Observation 1" and 1 in "Observation 2."
- $Count_{10}$: Count of all variables that are 1 in "Observation 1" and 0 in "Observation 2."
- $Count_{01}$: Count of all variables that are 0 in "Observation 1" and 1 in "Observation 2."
- $Count_{00}$: Count of all variables that are 0 in "Observation 1" and 0 in "Observation 2."

TABLE 5.3   Three Observations Measured over Five Binary Variables

	Variable 1	Variable 2	Variable 3	Variable 4	Variable 5
A	1	1	0	0	1
B	1	1	0	0	0
C	0	0	1	1	1

The following formula is used to calculate the Jaccard distance:

$$d = \frac{\text{Count}_{10} + \text{Count}_{01}}{\text{Count}_{11} + \text{Count}_{10} + \text{Count}_{01}}$$

The Jaccard distance is illustrated using Table 5.3.

Between observations A and B are two variables where both values are 1 (*Variable 1* and *Variable 2*), two values where both variables are 0 (*Variable 3* and *Variable 4*), one value where the value is 1 in observation A but 0 in observation B (*Variable 5*) and no values where a value is 0 in observation A and 1 in observation B. Therefore, the Jaccard distance between A and B is

$$d_{\text{A-B}} = (1 + 0)/(2 + 1 + 0) = 0.33$$

The Jaccard distance between A and C is

$$d_{\text{A-C}} = (2 + 2)/(1 + 2 + 2) = 0.8$$

The Jaccard distance between B and C is

$$d_{\text{B-C}} = (2 + 3)/(0 + 2 + 3) = 1.0$$

The Euclidean and Jaccard distance metrics are two examples of techniques for determining the distance between observations. Other techniques include Mahalanobis, City Block, Minkowski, Cosine, Spearman, Hamming, and Chebyshev (see the Further Reading section for references on these and other methods).

5.2.3   Agglomerative Hierarchical Clustering

Agglomerative hierarchical clustering is an example of a hierarchical method for grouping observations. It uses a "bottom-up" approach to clustering as it starts with each observation and progressively creates clusters

TABLE 5.4 Table of Observations to Cluster

Name	Variable 1	Variable 2	Variable 3	Variable 4	Variable 5
A	7.9	8.6	4.4	5.0	2.5
B	6.8	8.2	5.2	4.2	2.2
C	8.7	9.6	7.5	8.9	9.8
D	6.1	7.3	7.9	7.3	8.3
E	1.5	2.0	5.1	3.6	4.2
F	3.7	4.3	5.4	3.3	5.8
G	7.2	8.5	8.6	6.7	6.1
H	8.5	9.7	6.3	5.2	5.0
I	2.0	3.4	5.8	6.1	5.6
J	1.3	2.6	4.2	4.5	2.1
K	3.4	2.9	6.5	5.9	7.4
L	2.3	5.3	6.2	8.3	9.9
M	3.8	5.5	4.6	6.7	3.3
N	3.2	5.9	5.2	6.2	3.7

by merging observations together until all are a member of a final single cluster. The major limitation of agglomerative hierarchical clustering is that it is normally limited to data sets with fewer than 10,000 observations because the computational cost to generate the hierarchical tree can be high, especially for larger numbers of observations.

To illustrate the process of agglomerative hierarchical clustering, the data set of 14 observations measured over 5 variables as shown in Table 5.4 will be used. In this example, the variables are all measured on the same scale; however, where variables are measured on different scales they should be normalized to a comparable range (e.g., 0–1) prior to clustering. This prevents one or more variables from having a disproportionate weight and creating a bias in the analysis.

First, the distance between all pairs of observations is calculated. The method for calculating the distance along with the variables to include in the calculation should be set prior to clustering. In this example, we will use the Euclidean distance across all continuous variables shown in Table 5.4. The distances between all combinations of observations are summarized in a *distance matrix*, as illustrated in Table 5.5. In this example, the distances between four observations are shown (A, B, C, D) and each value in the table shows the distance between two indexed observations. The diagonal values are excluded, since these pairs are of the same observation. It should be noted that a distance matrix is usually symmetrical

TABLE 5.5 Distance Matrix Format

	A	B	C	D	...
A		$d_{A,B}$	$d_{A,C}$	$d_{A,D}$	...
B	$d_{B,A}$		$d_{B,C}$	$d_{B,D}$	...
C	$d_{C,A}$	$d_{C,B}$		$d_{C,D}$	...
D	$d_{D,A}$	$d_{D,B}$	$d_{D,C}$		...
...	...	...	...	...	

about the diagonal as the distance between, for example, A and B is the same as the distance between B and A.

For the 14 observations in Table 5.4, the complete initial distance matrix is shown in Table 5.6. This table is symmetrical about the diagonal since, as described previously, the ordering of the pairs is irrelevant when using the Euclidean distance. The two closest observations are identified (M and N in this example) and are merged into a single cluster. These two observations from now on will be considered a single group.

Next, all observations (minus the two that have been merged into a cluster) along with the newly created cluster are compared to see which observation or cluster should be joined into the next cluster. Since we are now analyzing individual observations and clusters, a *joining or linkage rule* is needed to determine the distance between an observation and a cluster of observations. This joining/linkage rule should be set prior to clustering. In Figure 5.8, two clusters have already been identified: Cluster A and Cluster B. We now wish to determine the distance between observation X and the cluster A or B (to determine whether or not to merge X with one of the two clusters). There are a number of ways to calculate the distance between an observation and an already established cluster including *average linkage*, *single linkage*, and *complete linkage*. These alternatives are illustrated in Figure 5.9.

- **Average linkage**: the distance between all members of the cluster (e.g., a, b, and c) and the observation under consideration (e.g., x) are calculated and the average is used for the overall distance.
- **Single linkage**: the distance between all members of the cluster (e.g., a, b, and c) and the observation under consideration (e.g., x) are calculated and the smallest is selected.
- **Complete linkage**: the distance between all members of the cluster (e.g., a, b, and c) and the observation under consideration (e.g., x) are calculated and the highest is selected.

TABLE 5.6 Calculated Distances Between All Pairs of Observations

	A	B	C	D	E	F	G	H	I	J	K	L	M	N
A		0.282	1.373	1.2	1.272	0.978	1.106	0.563	1.178	1.189	1.251	1.473	0.757	0.793
B	0.282		1.423	1.147	1.113	0.82	1.025	0.56	1.064	1.065	1.144	1.44	0.724	0.7
C	1.373	1.423		0.582	1.905	1.555	0.709	0.943	1.468	1.995	1.305	1.076	1.416	1.378
D	1.2	1.147	0.582		1.406	1.092	0.403	0.808	0.978	1.543	0.797	0.744	1.065	0.974
E	1.272	1.113	1.905	1.406		0.476	1.518	1.435	0.542	0.383	0.719	1.223	0.797	0.727
F	0.978	0.82	1.555	1.092	0.476		1.191	1.039	0.57	0.706	0.595	1.076	0.727	0.624
G	1.106	1.025	0.709	0.403	1.518	1.191		0.648	1.163	1.624	1.033	1.108	1.148	1.051
H	0.563	0.56	0.943	0.808	1.435	1.039	0.648		1.218	1.475	1.169	1.315	0.984	0.937
I	1.178	1.064	1.468	0.978	0.542	0.57	1.163	1.218		0.659	0.346	0.727	0.553	0.458
J	1.189	1.065	1.995	1.543	0.383	0.706	1.624	1.475	0.659		0.937	1.344	0.665	0.659
K	1.251	1.144	1.305	0.797	0.719	0.595	1.033	1.169	0.346	0.937		0.64	0.774	0.683
L	1.473	1.44	1.076	0.744	1.223	1.076	1.108	1.315	0.727	1.344	0.64		0.985	0.919
M	0.757	0.724	1.416	1.065	0.797	0.727	1.148	0.984	0.553	0.665	0.774	0.985		0.196
N	0.793	0.7	1.378	0.974	0.727	0.624	1.051	0.937	0.458	0.659	0.683	0.919	0.196	

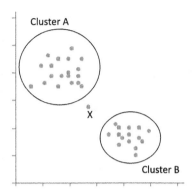

FIGURE 5.8 Comparing observation X with two clusters A and B.

Distances between all combinations of groups and observations are considered and the smallest distance is selected. Since we now may need to consider the distance between two clusters, the linkage/joining concept is extended to the joining of two clusters, as illustrated in Figure 5.10. The process of assessing all pairs of observations/clusters, then combining the pair with the smallest distance is repeated until there are no more clusters or observations to join together since only a single cluster remains.

Figure 5.11 illustrates this process for some steps based on the observations shown in Table 5.6. In step 1, it is determined that observations M and N are the closest and they are combined into a cluster, as shown. The horizontal length of the lines joining M and N reflects the distance at which the cluster was formed (0.196). From now on M and N will not be considered individually, but only as a cluster. In step 2, distances between all observations (except M and N) as well as the cluster containing M and N are calculated. To determine the distance between the individual observations and the cluster containing M and N, the average linkage rule was used. It is now determined that A and B should be joined as shown. Once again, all distances between the remaining ungrouped observations and the newly created clusters are calculated and the smallest distance selected.

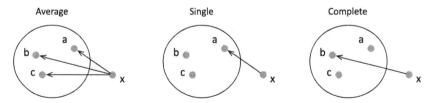

FIGURE 5.9 Different linkage rules for considering an observation and a cluster.

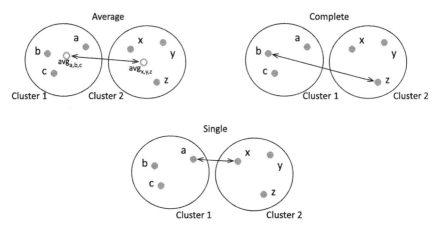

FIGURE 5.10   Different linkage rules for considering two clusters.

Steps 3 to 5 follow the same process. In step 6, the shortest distance is between observation I and the cluster containing M and N and in step 8, the shortest distance is between the cluster {M,N} and the cluster {F,I,K}. This process continues until only one cluster containing all the observations remains. Figure 5.12 shows the completed hierarchical clustering for all 14 observations.

When clustering completes, a tree called a *dendrogram* is generated showing the similarity between observations and clusters as shown in Figure 5.12. To divide a data set into a series of distinct clusters, we must select a distance at which the clusters are to be created. Where this distance intersects with a horizontal line on the tree, a cluster is formed, as illustrated

Step 1	Step 2	Step 3	Step 4	Step 4	Step 6	Step 7	Step 8
L	L	L	L	L	L	L	L
C	C	C	C	C	C	C	C
D	D	D	D	D	D	D	D
G	G	G	G	G	G	G	G
H	H	H	H	H	H	H	H
A	A	A	A	A	A	A	A
B	B	B	B	B	B	B	B
E	E	E	E	E	E	E	E
J	J	J	J	J	J	J	J
M	M	M	M	M	M	M	M
N	N	N	N	N	N	N	N
F	F	F	F	F	F	F	F
I	I	I	I	I	I	I	I
K	K	K	K	K	K	K	K

FIGURE 5.11   Steps 1 through 8 of the clustering process.

Distance

FIGURE 5.12   Completed hierarchical clustering for the 14 observations.

in Figure 5.13. In this example, three different distances (i, j, k) are used to divide the tree into clusters. Where this vertical line intersects with the tree (shown by the circles) at distance i, two clusters are formed: {L,C,D,G} and {H,A,B,E,J,M,N,F,I,K}; at distance j, four clusters are formed: {L}, {C,D,G}, {H,A,B}, and {E,J,M,N,F,I,K}; and at distance k, nine clusters are formed: {L}, {C}, {D,G}, {H}, {A,B}, {E,J}, {M,N}, {F}, and {I,K}. As illustrated in Figure 5.13, adjusting the cut-off distance will change the number of clusters created. Distance cut-offs toward the left will result in fewer clusters with more diverse observations within each cluster. Cut-offs toward the right will result in a greater number of clusters with more similar observations within each cluster.

Different joining/linkage rules change how the final hierarchical clustering is presented. Figure 5.14 shows the hierarchical clustering of the same set of observations using the average linkage, single linkage, and complete linkage rules. Since the barrier for merging observations and clusters is lowest with the single linkage approach, the clustering dendrogram may contain chains of clusters as well as clusters that are spread out. The barrier to joining clusters is highest with complete linkage; however, it is possible that an observation is closer to observations in other clusters than the cluster to which it has been assigned. The average linkage approach moderates the tendencies of the single or complete linkage approaches.

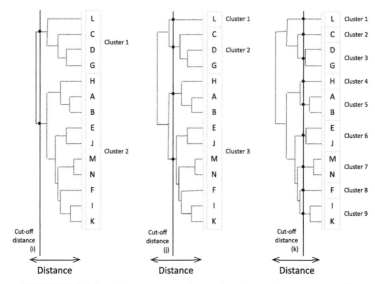

FIGURE 5.13   Cluster generation using three distance cut-offs.

The following example uses a data set of 392 cars that will be explored using hierarchical agglomerative clustering. A portion of the data table is shown in Table 5.7.

This data set was clustered using the *Euclidean* distance method and the *complete* linkage rule. The following variables were used in the clustering: *Displacement*, *Horsepower*, *Acceleration*, and *MPG* (miles per gallon).

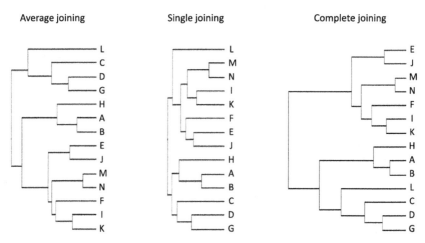

FIGURE 5.14   Different results using three different methods for joining clusters.

TABLE 5.7 Data Table Containing Automobile Observations

Car Name	MPG	Cylinders	Displacement	Horsepower	Weight	Acceleration	Model Year	Origin
Chevrolet Chevelle Malibu	18	8	307	130	3,504	12	70	American
Buick Skylark 320	15	8	350	165	3,693	11.5	70	American
Plymouth Satellite	18	8	318	150	3,436	11	70	American
Amc rebel sst	16	8	304	150	3,433	12	70	American
Ford Torino	17	8	302	140	3,449	10.5	70	American
Ford Galaxie 500	15	8	429	198	4,341	10	70	American

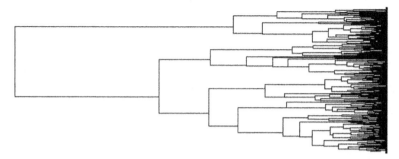

FIGURE 5.15 Hierarchical agglomerative clustering dendrogram generated for the automobile data set.

The dendrogram in Figure 5.15 of the generated clusters shows the relationships between observations based on the similarity of the four selected variables. Each horizontal line at the right represents a single automobile and the order of the observations is related to how similar each car is to its neighbors.

In Figure 5.16 a distance cut-off has been set such that the data is divided into three clusters. In addition to showing the dendrogram, three small multiple charts illustrate the composition of each cluster – the highlighted region – for each of the four variables. Cluster 1 is a set of 97 observations with low fuel efficiency and low acceleration values, and generally higher values for horsepower and displacement. Cluster 2 contains 85 observations with generally good fuel efficiency and acceleration as well as low horsepower and displacement. Cluster 3 contains 210 observations, the majority of which have average fuel efficiency and acceleration as well as few high values for displacement or horsepower.

To explore the data set further we can adjust the distance cut-off to generate different numbers of clusters. Figure 5.17 displays the case in which the distance was set to create nine clusters. Cluster 1 (from Figure 5.16) is now divided into three clusters of sizes 36, 4, and 57. The new cluster 1 is a set of 36 observations with high horsepower and displacement values, as well as low fuel efficiency and acceleration; cluster 2 represents a set of only 4 cars with the worst fuel efficiency and improved acceleration; and cluster 3 is a set of 57 cars with lower horsepower than cluster 1 or 2 and improved MPG values. Similarly, cluster 2 (from Figure 5.16) is now divided into three clusters of sizes 33, 4, and 48 and cluster 3 (from Figure 5.16) is now divided into three clusters of sizes 73, 35, and 102.

Since the ordering of the observations provides insight into the data set's organization, a clustering dendrogram is often accompanied by a colored

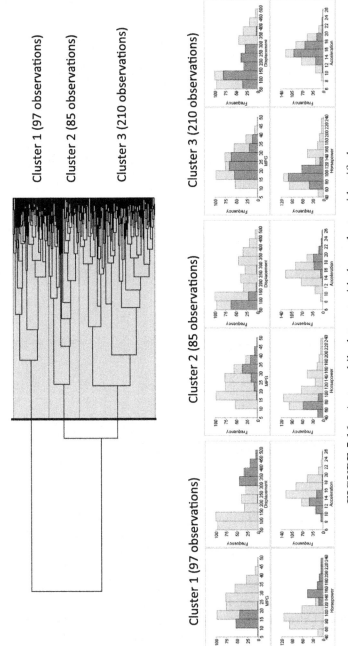

FIGURE 5.16 Automobile data set with three clusters identified.

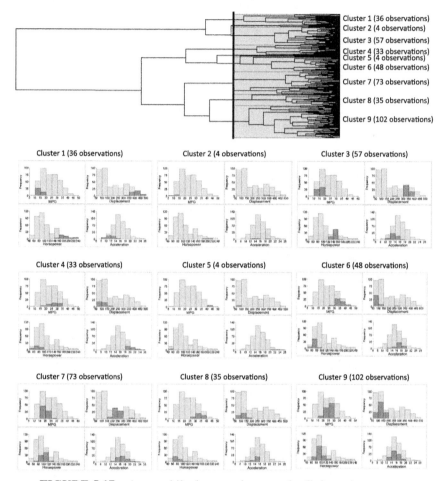

FIGURE 5.17   Automobile data set cluster and split into nine groups.

heatmap that uses different colors or shades to represent the different observation values across variables of interest. For example, in Figure 5.18 the 14 observations from Table 5.6 have been clustered using agglomerative hierarchical cluster, based on the Euclidean distance over the 5 continuous variables and the average linkage joining method. A heatmap, shown to the right of the dendrogram, is used to represent the data values for the 14 observations. Different shades of gray are used to represent the data values (binned as shown in the legend). It is possible to see patterns in the data set using this approach because observations with similar patterns have been grouped close together.

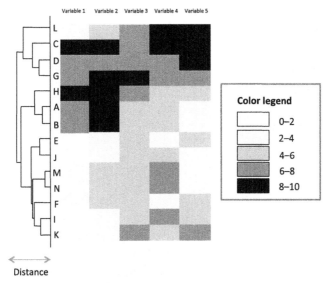

FIGURE 5.18   Clustering dendrogram coupled with a colored heatmap.

5.2.4 *k*-Means Clustering

k-Means clustering is an example of a nonhierarchical method for grouping a data set. It groups data using a "top-down" approach since it starts with a predefined number of clusters and assigns all observations to each of them. There are no overlaps in the groups; each observation is assigned only to a single group. This approach is computationally faster and can handle greater numbers of observations than agglomerative hierarchical clustering. However, there are several disadvantages to using this method. The most significant is that the number of groups must be specified before creating the clusters and this number is not guaranteed to provide the best partitioning of the observations. Another disadvantage is that when a data set contains many outliers, *k*-means may not create an optimal grouping (discussed later in this section). Finally, no hierarchical organization is generated using *k*-means clustering and hence there is no ordering of the individual observations.

The process of generating clusters starts by defining the value *k*, which is the number of groups to create. The method then initially allocates an observation – usually selected randomly – to each of these groups. Next, all other observations are compared to each of the allocated observations and placed in the group to which they are most similar. The center point for each of these groups is then calculated. The grouping process continues by

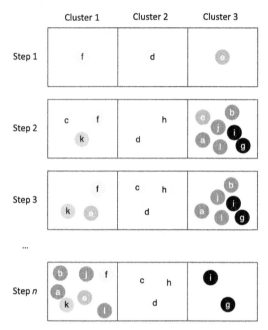

FIGURE 5.19   Illustrating conceptually k-means clustering.

determining the distance from all observations to these new group centers. If an observation is closer to the center of another group, it is moved to the group that it is closest to. The centers of its old and new groups are then recalculated. The process of comparing and moving observations where appropriate is repeated until no observations are moved after a recalculation of the group's center.

To illustrate the process of clustering using k-means, a set of 12 hypothetical observations are used: a, b, c, d, e, f, g, h, i, j, k, and l. These observations are shown as colored circles in Figure 5.19. It was determined at the start that three groups should be generated. Initially, an observation is randomly assigned to each of the three groups as shown in step 1: f to cluster 1, d to cluster 2, and e to cluster 3. Next, all remaining observations are assigned to the cluster to which they are closest using one of the distance functions described earlier. For example, observation c is assigned to cluster 1 since it is closer to f than to d or e. Once all observations have been assigned to an initial cluster, the point at the center of each cluster is then calculated. Next, distances from each observation to the center of each cluster are calculated. It is determined in step 3 that observation c is closer to the center of cluster 2 than the other two clusters, so c is moved to cluster 1.

TABLE 5.8 Data Table Used in the *k*-Mean Clustering Example

Name	Variable 1	Variable 2	Variable 3	Variable 4	Variable 5
A	7.9	8.6	4.4	5.0	2.5
B	6.8	8.2	5.2	4.2	2.2
C	8.7	9.6	7.5	8.9	9.8
D	6.1	7.3	7.9	7.3	8.3
E	1.5	2.0	5.1	3.6	4.2
F	3.7	4.3	5.4	3.3	5.8
G	7.2	8.5	8.6	6.7	6.1
H	8.5	9.7	6.3	5.2	5.0
I	2.0	3.4	5.8	6.1	5.6
J	1.3	2.6	4.2	4.5	2.1
K	3.4	2.9	6.5	5.9	7.4
L	2.3	5.3	6.2	8.3	9.9
M	3.8	5.5	4.6	6.7	3.3
N	3.2	5.9	5.2	6.2	3.7

It is also determined that e and k are now closer to the center of cluster 1 so these observations are moved to cluster 1. Since the contents of all three clusters have changed, the centers for all clusters are recalculated. This process continues until no more observations are moved between clusters, as shown in step *n* on the diagram.

A disadvantage of *k*-means clustering is that when a data set contains many outliers, *k*-means may not create an optimal grouping. This is because the reassignment of observations is based on closeness to the center of the cluster and outliers pull the centers of the clusters in their direction, resulting in assignment of the remaining observations to other groups.

The following example will illustrate the process of calculating the center of a cluster. The observations in Table 5.8 are grouped into three clusters using the *Euclidean* distance to determine the distance between observations. A single observation is randomly assigned to the three clusters as shown in Figure 5.20: I to cluster 1, G to cluster 2, D to cluster 3. All other

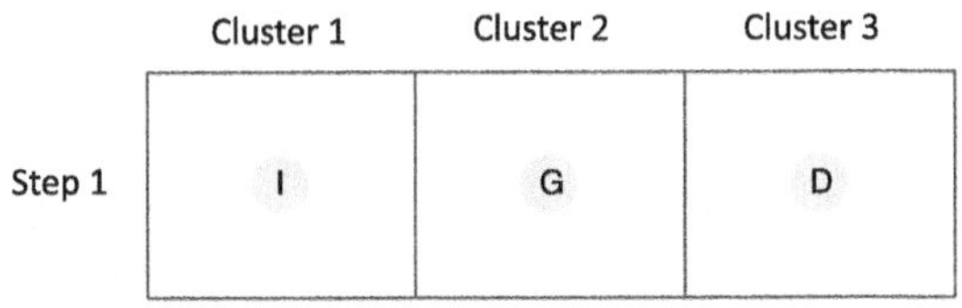

FIGURE 5.20 Single observation randomly assigned to each cluster.

TABLE 5.9 **Euclidean Distances and Cluster Assignments**

Name	Cluster 1	Cluster 2	Cluster 3	Cluster Assignment
A	1.178	1.106	1.2	2
B	1.064	0.025	1.147	2
C	1.468	0.709	0.582	3
E	0.542	1.518	1.406	1
F	0.57	1.191	1.092	1
H	1.218	0.648	0.808	2
J	0.659	1.624	1.543	1
K	0.346	1.033	0.797	1
L	0.727	1.108	0.744	1
M	0.553	1.148	1.065	1
N	0.458	1.051	0.974	1

observations are compared to the three clusters by calculating the distance between the observations and I, G, and D. Table 5.9 shows the Euclidean distance to I, G, and D from every other observation, along with the cluster it is initially assigned to. All observations are now assigned to one of the three clusters (Figure 5.21).

Next, the center of each cluster is calculated by taking the average value for each variable in the group, as shown in Table 5.10. For example, the center of cluster 1 is now

$$\{\text{Variable } 1 = 2.65; \text{ Variable } 2 = 3.99; \text{ Variable } 3 = 5.38;$$
$$\text{Variable } 4 = 5.58; \text{ Variable } 5 = 5.25\}$$

Each observation is now compared to the centers of each cluster and the process of examining the observations and moving them as appropriate is repeated until no further moves are needed. In this example, the final assignment is shown in Figure 5.22.

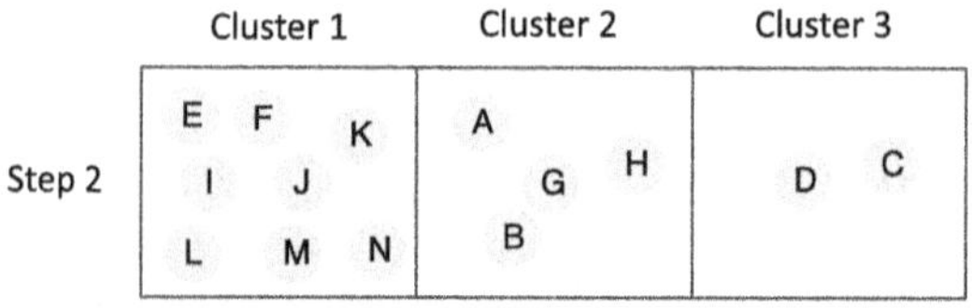

FIGURE 5.21 Initial assignment of other observations to each cluster.

TABLE 5.10   Calculating the Center of Each Cluster

Cluster 1

Name	Variable 1	Variable 2	Variable 3	Variable 4	Variable 5
E	1.5	2	5.1	3.6	4.2
F	3.7	4.3	5.4	3.3	5.8
I	2	3.4	5.8	6.1	5.6
J	1.3	2.6	4.2	4.5	2.1
K	3.4	2.9	6.5	5.9	7.4
L	2.3	5.3	6.2	8.3	9.9
M	3.8	5.5	4.6	6.7	3.3
N	3.2	5.9	5.2	6.2	3.7
Average (Center)	2.65	3.99	5.38	5.58	5.25

Cluster 2

Name	Variable 1	Variable 2	Variable 3	Variable 4	Variable 5
A	7.9	8.6	4.4	5	2.5
B	6.8	8.2	5.2	4.2	2.2
G	7.2	8.5	8.6	6.7	6.1
H	8.5	9.7	6.3	5.2	5
Average (Center)	7.60	8.75	6.13	5.28	3.95

Cluster 3

Name	Variable 1	Variable 2	Variable 3	Variable 4	Variable 5
C	8.7	9.6	7.5	8.9	9.8
D	6.1	7.3	7.9	7.3	8.3
Average (Center)	7.40	8.45	7.70	8.10	9.05

A data set of 392 cars is grouped using k-means clustering. This is the same data set used in the agglomerative hierarchical clustering example. The Euclidean distance was used and the number of clusters was set to 3. The same set of variables was used as in the agglomerative hierarchical clustering example. Although both methods produce similar results, they are not identical. k-means cluster 3 (shown in Figure 5.23) is almost

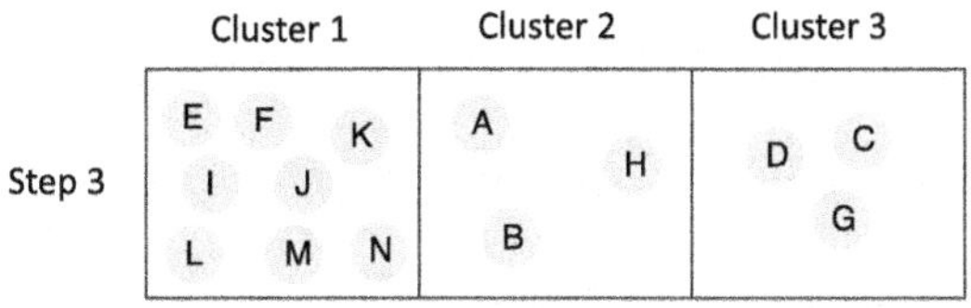

FIGURE 5.22   Final assignment of observations using k-means clustering.

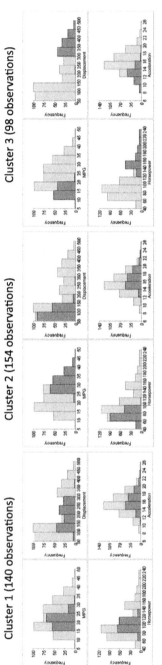

FIGURE 5.23 Three clusters generated using *k*-means clustering.

identical to the agglomerative hierarchical cluster 1 (shown in Figure 5.16). There is some similarity between cluster 2 (k-means) and cluster 3 (agglomerative hierarchical) as well as cluster 1 (k-means) and cluster 3 (agglomerative hierarchical). Figure 5.24 shows the results of a k-means method generating nine groups.

5.3 ASSOCIATION RULES

5.3.1 Overview

The *association rules* method groups observations and attempts to discover links or associations between different attributes of the group. Associative rules can be applied in many situations such as data mining retail transactions. This method generates rules from the groups as, for example,

IF the customer is age 18 AND

the customer buys paper AND

the customer buys a hole punch

THEN the customer buys a binder

The rule states that 18-year-old customers who purchase paper and a hole punch often buy a binder at the same time. Using this approach, the rule would be generated directly from a data set and using this information the retailer may decide, for example, to create a package of products for college students.

The association rules method is an example of an unsupervised grouping method. (Recall from Section 5.2 that unsupervised methods are undirected, i.e., no specific variable is selected to guide the process.) The advantages of this method include the generation of rules that are easy to understand, the ability to perform an action based on the rule as in the previous example which allowed the retailer to apply the rule to make changes to the marketing strategy, and the possibility of using this technique with large numbers of observations.

There are, however, limitations. This method forces you to either restrict your analysis to variables that are categorical or convert continuous variables to categorical variables. Generating the rules can be computationally expensive, especially where a data set has many variables or many possible values per variable, or both. There are ways to make the analysis run faster but they often compromise the final results. Finally, this method can generate large numbers of rules that must be prioritized and interpreted.

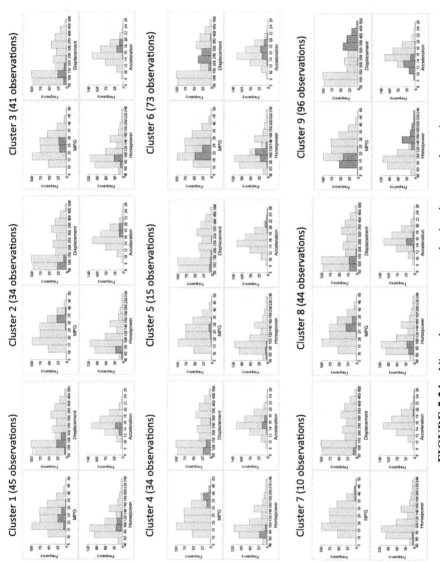

FIGURE 5.24 Nine clusters generated using *k*-means clustering.

TABLE 5.11   Data Table of Three Sample Examples Observations with Three Variables

Customer ID	Gender	Purchase
932085	Male	Television
596720	Female	Camera
267375	Female	Television

In this method, creating useful rules from the data is done by grouping it, extracting rules from the groups, and then prioritizing the rules. The following sections describe the process of generating association rules.

5.3.2 Grouping by Combinations of Values

Let us first consider a simple situation concerning a shop that sells only cameras and televisions. A data set of 31,612 sales transactions is used which contains three variables: *Customer ID*, *Gender*, and *Purchase*. The variable *Gender* identifies whether the buyer is "male" or "female". The variable *Purchase* refers to the item purchased and can have only two values: "camera" and "television." Table 5.11 shows three rows from this table. Each row of the table contains an entry where values are assigned for each variable (i.e., there are no missing values). By grouping this set of 31,612 observations, based on specific values for the variables *Gender* and *Purchase*, the groups shown in Table 5.12 are generated. There are eight ways of grouping this trivial example based on the values for the different categories of *Gender* and *Purchase*. For example, there are 7,889 observations where *Gender* is "male" and *Purchase* is "camera."

TABLE 5.12   Grouping by Different Value Combinations

Group Number	Count	Gender	Purchase
Group 1	16,099	Male	Camera or Television
Group 2	15,513	Female	Camera or Television
Group 3	16,106	Male or Female	Camera
Group 4	15,506	Male or Female	Television
Group 5	7,889	Male	Camera
Group 6	8,210	Male	Television
Group 7	8,217	Female	Camera
Group 8	7,296	Female	Television

TABLE 5.13   Table Showing Groups by Different Value Combinations

Group Number	Count	Gender	Purchase	Income
Group 1	16,099	Male	Camera or Television	Below $50K or Above $50K
Group 2	15,513	Female	Camera or Television	Below $50K or Above $50K
Group 3	16,106	Male or Female	Camera	Below $50K or Above $50K
Group 4	15,506	Male or Female	Television	Below $50K or Above $50K
Group 5	15,854	Male or Female	Camera or Television	Below $50K
Group 6	15,758	Male or Female	Camera or Television	Above $50K
Group 7	7,889	Male	Camera	Below $50K or Above $50K
Group 8	8,210	Male	Television	Below $50K or Above $50K
Group 9	8,549	Male	Camera or Television	Below $50K
Group 10	7,550	Male	Camera or Television	Above $50K
Group 11	8,217	Female	Camera	Below $50K or Above $50K
Group 12	7,296	Female	Television	Below $50K or Above $50K
Group 13	7,305	Female	Camera or Television	Below $50K
Group 14	8,208	Female	Camera or Television	Above $50K
Group 15	8,534	Male or Female	Camera	Below $50K
Group 16	7,572	Male or Female	Camera	Above $50K
Group 17	7,320	Male or Female	Television	Below $50K
Group 18	8,186	Male or Female	Television	Above $50K
Group 19	4,371	Male	Camera	Below $50K
Group 20	3,518	Male	Camera	Above $50K
Group 21	4,178	Male	Television	Below $50K
Group 22	4,032	Male	Television	Above $50K
Group 23	4,163	Female	Camera	Below $50K
Group 24	4,054	Female	Camera	Above $50K
Group 25	3,142	Female	Television	Below $50K
Group 26	4,154	Female	Television	Above $50K

If an additional variable is added to this data set, the number of possible groups will increase as, for example, if another variable *Income* which has two values – above $50K and below $50K – is added to the table, the number of groups would increase to 26 as shown in Table 5.13.

Increasing the number of variables or the number of possible values for each variable or both increases the number of groups. When the number of

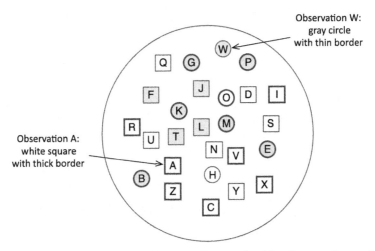

FIGURE 5.25   Twenty-six observations characterized by shape, color, and border attributes.

groups becomes too large, it becomes impractical to generate all combinations. However, most data sets contain many combinations of values with only a few or no observations. Techniques for generating the groups take advantage of this by requiring that groups reach a certain size before they are generated. This results in fewer groups and shortens the time required to compute the results. However, care should be taken in setting this cut-off value since rules can only be created from groups that are generated. For example, if this number is set to 10, then no rules will be generated from groups containing fewer than 10 observations. Subject matter knowledge and information generated from the data characterization phase will help in setting the cut-off value. There is a trade-off between speed of computation and how fine-grained the rules need to be (i.e., rules based on a few observations).

5.3.3 Extracting and Assessing Rules

So far a data set has been grouped according to specific values for each of the variables. In Figure 5.25, 26 observations (A through Z) are characterized by three variables: *Shape*, *Color*, and *Border*. Observation A has *Shape* = square, *Color* = white and *Border* = thick and observation W has *Shape* = circle, *Color* = gray and *Border* = thin.

As described in the previous section, the observations are grouped. An example of such a grouping is shown in Figure 5.26 where *Shape* = circle, *Color* = gray and *Border* = thick.

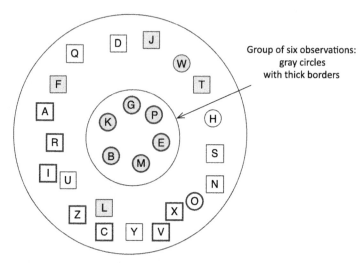

FIGURE 5.26   Group of six observations (gray circles with thick borders).

The next step is to extract a rule from the group. There are three possible rules (containing all three variables) that could be derived from this group (Figure 5.26):

Rule 1

IF *Color* = gray AND
Shape = circle
THEN *Border* = thick

Rule 2

IF *Border* = thick AND
Color = gray
THEN *Shape* = circle

Rule 3

IF *Border* = thick AND
Shape = circle
THEN *Color* = gray

We now examine each rule in detail and make a comparison to the whole data set in order to prioritize the rules. Three values are calculated to support this assessment: *support*, *confidence*, and *lift*.

Support   The *support* is a measure of the number of observations a rule maps on to. Its value is the proportion of the observations a rule selects out of all observations in the data set. In this example, the data set has 26 observations and the group of gray circles with a thick border is 6, then the group has a support value of 6 out of 26 or 0.23 (23%).

Confidence   Each rule is divided into two parts: antecedent and consequence. The IF-part or *antecedent* refers to a list of statements linked with AND in the first part of the rule. For example,

IF *Color* = gray AND
Shape = circle
THEN *Border* = thick

The IF-part is the list of statements *Color* = gray AND *Shape* = circle. The THEN-part of the rule or *consequence* refers to statements after the THEN (*Border* = thick in this example).

The *confidence score* is a measure for how predictable a rule is. The confidence or predictability value is calculated using the support for the entire group divided by the support for all observations satisfied by the IF-part of the rule:

Confidence = group support/IF-part support

For example, the confidence value for Rule 1 is calculated using the support value for the group and the support value for the IF-part of the rule (see Figure 5.27).

Rule 1

IF *Color* = gray AND
Shape = circle
THEN *Border* = thick

The support value for the group (gray circles with a thick border) is 0.23 and the support value for the IF-part of the rule (gray circles) is 7 out of 26 or 0.27. To calculate the confidence, we divide the support for the group by the support for the IF-part:

Confidence = 0.23/0.27 = 0.85

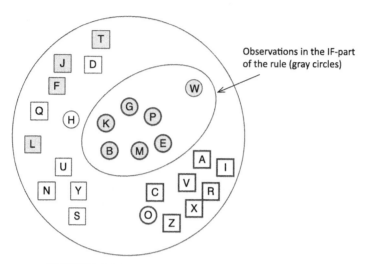

FIGURE 5.27   Seven observations for gray circles.

Confidence values range from no confidence (0) to high confidence (1). Since a value of 0.85 is close to 1, we have a high degree of confidence in this rule.

Lift   The confidence value does not indicate the strength of the association between gray circles (IF-part) and thick borders (THEN-part). The lift score takes this into account. The *lift* is often described as the importance of the rule as it describes the association between the IF-part of the rule and the THEN-part of the rule. It is calculated by dividing the confidence value by the support value across all observations of the THEN-part:

$$\text{Lift} = \text{confidence}/\text{THEN-part support}$$

For example, the lift for Rule 1:

Rule 1

IF *Color* = gray AND
Shape = circle
THEN *Border* = thick

is calculated using the confidence and the support for the THEN-part of the rule (see Figure 5.28). The confidence for Rule 1 is calculated as 0.85 and the support for the THEN-part of the rule (thick borders) is 14 out of 26 or

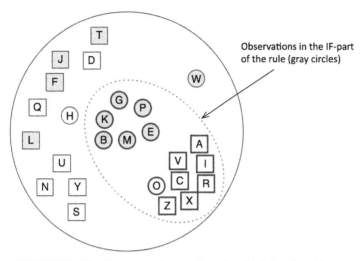

Observations in the IF-part
of the rule (gray circles)

FIGURE 5.28   Fourteen observations for thick border objects.

0.54. To calculate the lift value, the confidence is divided by the support value for the THEN-part of the rule:

$$Lift = 0.85/0.54 = 1.57$$

Lift values greater than 1 indicate a positive association.

Figure 5.29 is used to determine the confidence and support for all three potential rules:

The following shows the calculations for support, confidence, and lift for the three rules:

Rule 1

Support $= 6/26 = 0.23$
Confidence $= 0.23/(7/26) = 0.85$
Lift $= 0.85/(14/26) = 1.58$

Rule 2

Support $= 6/26 = 0.23$
Confidence $= 0.23/(6/26) = 1$
Lift $= 1/(9/26) = 2.89$

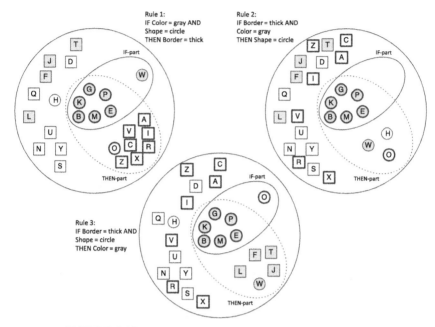

FIGURE 5.29 Separating objects for each rule calculation.

Rule 3

Support $= 6/26 = 0.23$

Confidence $= 0.23/(7/26) = 0.85$

Lift $= 0.85/(11/26) = 2.01$

The values are summarized in Table 5.14.

Rule 2 would be considered the most interesting because it has a confidence score of 1 and a high positive lift score indicating that gray shapes with a thick border are likely to be circles.

TABLE 5.14 Summary of Support, Confidence, and Lift for the Three Rules

	Rule 1	Rule 2	Rule 3
Support	0.23	0.23	0.23
Confidence	0.85	1.0	0.85
Lift	1.58	2.89	2.01

5.3.4 Example

In this example, we will compare two rules generated from the Adult data set available from Bache and Lichman (2013), a set of income data that includes the following variables along with all possible values shown in parenthesis:

- *Class of work* (Private, Self-emp-not-inc, Self-emp-inc, Federal-gov, Local-gov, State-gov, Without-pay, Never-worked)
- *Education* (Bachelors, Some-college, 11th, HS-grad, Prof-school, Assoc-acdm, Assoc-voc, 9th, 7th–8th, 12th, Masters, 1st–4th, 10th, Doctorate, 5th–6th, Preschool)
- *Income* (>50K, ≤50K)

There are 32,561 observations. Using the associative rule method, many rules were identified. For example,

Rule 1

IF *Class of work* is Private AND
Education is Doctorate
THEN *Income* is <=50K

Rule 2

IF *Class of work* is Private AND
Education is Doctorate
THEN *Income* is >50K

Here is a summary of the counts:
Class of work is Private: 22,696 observations
Education is Doctorate: 413 observations
Class of work is private and *Education* is Doctorate: 181 observations
Income is <=50K: 24,720 observations
Income is >50K: 7841 observations

Table 5.15 shows the information calculated for the rules. Of the 181 observations where *Class of work* is Private and *Education* is Doctorate, 132 (73%) of those observations also had *Income* >50K. This is reflected in the much higher confidence score for Rule 2 (0.73) compared to Rule 1 (0.27). Over the entire data set of 32,561 observations there are about three times the number of observations where income ≤50K as

TABLE 5.15   Summary of Scores for Two Rules

	Rule 1	Rule 2
Count	49	132
Support	0.0015	0.0041
Confidence	0.27	0.73
Lift	0.36	3.03

compared to observations where the income is >50K. The lift term takes into consideration the relative frequency of the THEN-part of the rule. Hence, the lift value for Rule 2 is considerably higher (3.03) than the lift value for Rule 1. Rule 2 has good confidence and lift values, making it an interesting rule. Rule 1 has poor confidence and lift values. The following illustrates examples of other generated rules:

Rule 3

> IF *Class of work* is State-gov AND
> *Education* is 9th
> THEN *Income* is <=50K
> (Count: 6; Support: 0.00018; Confidence: 1; Lift: 1.32)

Rule 4

> IF *Class of work* is Self-emp-inc AND
> *Education* is Prof-school
> THEN *Income* is >50K
> (Count: 78; Support: 0.0024 Confidence: 0.96; Lift: 4)

Rule 5

> IF *Class of work* is Local-gov AND
> *Education* is 12th
> THEN *Income* is <=50K
> (Count: 17; Support: 0.00052; Confidence: 0.89; Lift: 1.18)

5.4   LEARNING DECISION TREES FROM DATA

5.4.1   Overview

It is often necessary to ask a series of questions before coming to a decision. The answers to one question may lead to other questions or may lead to

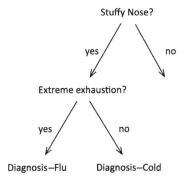

FIGURE 5.30   Decision tree for the diagnosis of colds and flu.

a decision. For example, you may visit a doctor and your doctor may ask you to describe your symptoms. You respond by saying you have a stuffy nose. In trying to diagnose your condition the doctor may ask you further questions such as whether you are suffering from extreme exhaustion. Answering yes may suggest you have the flu, whereas answering no might suggest that you have a cold. This line of questioning is common to many decision-making processes and can be shown visually as a decision tree, as shown in Figure 5.30.

Decision trees are often generated by hand to precisely and consistently define a decision-making process; however, they can also be generated automatically from the data. They consist of a series of decision points based on certain selected variables. Figure 5.31 illustrates a simple decision tree. This decision tree generated was based on a data set of cars that included variables for the number of cylinders (*Cylinders*) and the car's fuel efficiency (*MPG*). The decision tree uses the number of cylinders (*Cylinders*) to attempt to achieve the goal of classifying the observations according to their fuel efficiency. At the top of the tree is a node representing the entire data set of 392 observations (Size = 392). The data set is initially divided into two subsets: to the left is a set of 203 cars (i.e., Size = 203) where the number of cylinders is fewer than 5 and to the right are the remaining observations (number of cylinders 5 or greater). We describe in a later section how this division was determined. Cars with fewer than five cylinders are grouped together as they generally have good fuel efficiency. In this case the average of *MPG* is 29.11. The remaining 189 cars are further classified into a set of 86 cars where the number of cylinders is fewer than 7. This set does not include the cars with fewer than five cylinders because those cars were moved to a separate group in an earlier step. The set of 86 cars are grouped together as they generally have reasonable fuel efficiency—the average of *MPG* is 20.23—as compared with the poor fuel

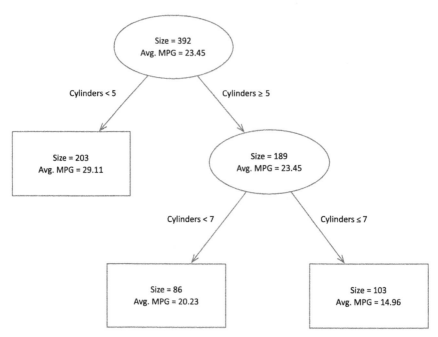

FIGURE 5.31   Decision tree generated from a data set of cars.

efficiency of the remaining group. The remaining group is a set of 103 cars where the number of cylinders in each car is greater than 7 and the average of *MPG* is 14.96.

In contrast with clustering or association rules, decision trees are an example of a *supervised* method. Supervised methods, as opposed to unsupervised methods, are an attempt to place (classify) each observation into interesting groups (based on a selected variable). These methods iterate over a training set of observations and adjust parameters as the classifier correctly or incorrectly classifies each observation. In this example, the data set was classified into groups using the variable *MPG* to guide how the tree was constructed. Figure 5.32 illustrates how the tree, guided by the data, was put together. A histogram of the *MPG* data is shown alongside the nodes used to classify the vehicles. The overall shape of the histograms depicts the frequency distribution for the *MPG* variable. The highlighted frequency distribution is the subset within the node. The frequency distribution for the node containing 203 observations shows a set biased toward good fuel efficiency, whereas for the node of 103 observations it illustrates a set biased toward poor fuel efficiency. The *MPG* variable has not been used in any of the decision points, only the number of cylinders. This is

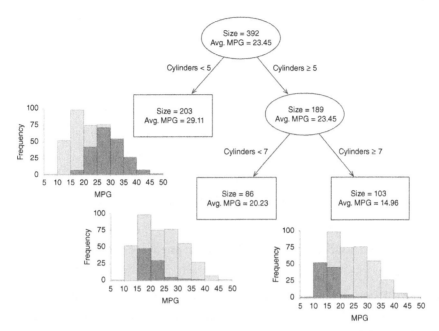

FIGURE 5.32   Decision tree illustrating the use of a response variable MPG to guide the tree generation.

a trivial example, but it shows how a data set can be divided into regions using decision trees.

There are two primary reasons to use decision trees. First, they are easy to understand and use in explaining how decisions are reached based on multiple criteria. Second, they can handle categorical and continuous variables since they partition a data set into distinct regions based on ranges or specific values. However there are disadvantages, as building decision trees can be computationally expensive, particularly when analyzing a large data set with many continuous variables. In addition, generating a useful decision tree automatically can be challenging, since large and complex trees can be easily generated; trees that are too small may not capture enough information; and generating the "best" tree through optimization is difficult. At the end of this chapter, there are a series of references to methods for optimizing decision trees further.

5.4.2 Splitting

A tree is made up of a series of decision points, where the split of the entire set of observations or a subset of the observations is based on some

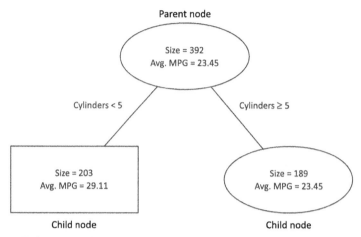

FIGURE 5.33   Relationship between parent and child nodes.

criteria. Each point in the tree represents a set of observations called a *node*. The relationship between two connected nodes is defined as a *parent–child* relationship. The larger set that will be divided into two or more smaller sets is the *parent* node. The nodes resulting from the division of the parent are *child* nodes as shown in Figure 5.33. A child node with no children is a *leaf* node as shown in Figure 5.34.

A table of data is used to generate a decision tree where certain variables are used as potential decision points (*splitting variables*) and one variable is used to guide the construction of the tree (*response variable*). The response variable will be used to guide which splitting variables are selected and at what value the split is made. A decision tree splits the data set into increasingly smaller, nonoverlapping subsets. The topmost node, or *root* of the tree, contains all observations. Based on some criteria, the observations are usually split into two new nodes, where each node represents a subset of observations as shown in Figure 5.35. Node N1 represents all observations. By analyzing all splitting variables and examining many splitting points for each variable, an initial split is made (C1). The data set represented at node N1 is now divided into a subset N2 that meets criteria C1, and a subset represented by node N3 that does not satisfy the criteria.

The process of examining the variables to determine a criterion for splitting is repeated for all subsequent nodes. Additionally, a condition is needed to end the process. For example, the process can stop when the size of the subset is less than a predetermined value. In Figure 5.36, each of the two newly created subsets (N2 and N3) is examined in turn to determine if they should be further split or whether the splitting should stop.

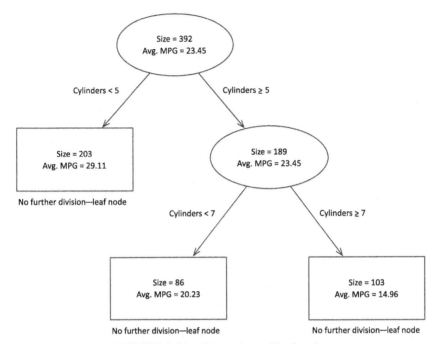

FIGURE 5.34   Illustration of leaf nodes.

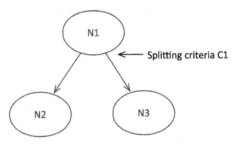

FIGURE 5.35   Node N1 split into two based on the criteria C1.

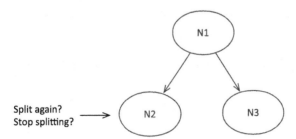

FIGURE 5.36   Evaluation of whether to continue to grow the tree.

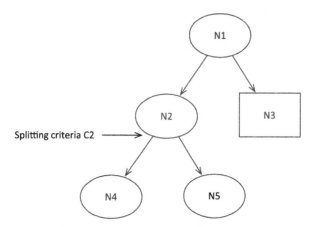

FIGURE 5.37 Tree further divided.

In Figure 5.37, the subset at node N2 is examined to determine if the splitting should stop. In this case, because the condition for stopping splitting is not met, the process continues. All the variables assigned as splitting variables are considered along with alternative values. The best criterion is selected and the set at node N2 is again divided into two subsets, represented by N4 and N5. Node N4 represents a set of observations that satisfies the splitting criteria (C2) and node N5 the remaining set of observations. Next, node N3 is examined. In this case, the condition to stop splitting is met and the process is halted.

5.4.3 Splitting Criteria

Dividing Observations It is common for the split at each level to be a two-way split. Although there are methods that split more than two ways, care should be taken when using these methods because making too many splits early in the construction of the tree may result in missing interesting relationships that become exposed as tree construction continues. This results from dividing the set into small groups based on a single criterion. Figure 5.38 illustrates the two alternatives.

Any variable type can be split using a two-way split (as shown in Figure 5.39):

- **Dichotomous**: Variables with two values are the most straightforward to split since each branch represents a specific value. For example, a variable *Temperature* may have only two values: "hot" and "cold."

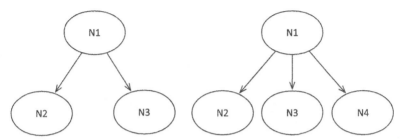

FIGURE 5.38   Alternative splitting of nodes.

Observations will be split to separate those with "hot" and those with "cold" temperature values.

• **Nominal**: Since nominal values are discrete values with no order, a two-way split is accomplished by one subset being composed of a set of observations that equal a certain value and the other being those observations that do not equal that value. For example, a variable *Color* that can take the values "red," "green," "blue," and "black" may be split two-ways. Observations, for example, which have *Color* equaling "red" generate one subset and those not equaling "red" creating the other subset, that is, "green," "blue," and "black."

• **Ordinal**: In the case where a variable's discrete values are ordered, the resulting subsets may be made up of more than one value, as

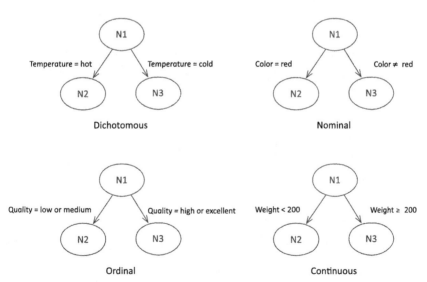

FIGURE 5.39   Splitting examples based on variable type.

long as the ordering is retained. For example, a variable *Quality* with possible values "low," "medium," "high," and "excellent" may be split four ways. For example, observations with *Quality* equaling "low" or "medium" may be in one subset and observations with *Quality* equaling "high" and "excellent" in another. Another possibility is that "low" values of *Quality* are in one set and "medium," "high," and "excellent" values are in the other set.

- **Continuous**: For variables with continuous values to be split two-ways, a specific cut-off value needs to be determined so that observations with values less than the cut-off are in the subset on the left and those with values greater than or equal to are in the subset on the right. For example, a variable *Weight* which can take any value between 0 and 1,000 with a selected cut-off of 200. The left subset would be those observations where the *Weight* is below 200 and the right subset those where the *Weight* is greater than or equal to 200.

A splitting criterion usually has two components: (1) the variable on which to split and (2) the values of that variable to use for the split. To determine the best split, a ranking is made of all possible splits of all variables using a score calculated for each split. There are many ways to rank the split. The following describes two approaches for prioritizing splits, based on whether the response is categorical or continuous.

Scoring Splits for Categorical Response Variables   To illustrate how to score splits when the response is a categorical variable, three splits (**Split a**, **Split b**, **Split c**) for a set of observations are shown in Figure 5.40. The objective for an optimal split is to create subsets which result in observations with a single response value. In this example, there are 20 observations prior to splitting. The response variable (*Temperature*) has

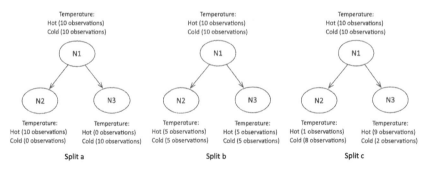

FIGURE 5.40   Evaluating splits based on categorical response data.

two possible values: "hot" and "cold." Prior to the split, the response has an even distribution: the number of observations where the *Temperature* equals "hot" is 10 and the number of observations where the *Temperature* equals "cold" is also 10.

Different criteria are considered for splitting these observations, which results in different distributions of the response variables for each subset (N2 and N3):

- **Split a**: Each subset contains 10 observations. All 10 observations in N2 have "hot" temperature values and all 10 observations in node N3 are "cold."
- **Split b**: Again each subset (N2 and N3) contain 10 observations. However, in this case there is an even distribution of "hot" and "cold" values in each subset.
- **Split c**: In this case the splitting criterion results in two subsets where node N2 has nine observations (1 "hot" and 8 "cold") and node N3 has 11 observations (9 "hot" and 2 "cold").

Split a is the best split since each node contains observations where the response for each node is all of the same category. **Split b** results in the same even split of "hot" and "cold" values (50% "hot," 50% "cold") in each of the resulting nodes (N2 and N3) and would not be considered a good split. **Split c** is a good split even though the split is not as clean as **Split a,** since both subsets have a mixture of "hot" and "cold" values. The proportion of "hot" and "cold" values in node N2 is biased toward cold values and in node N3 toward hot values. The "goodness" of the splitting criteria is determined by how clean each split is: it is based on the proportion of the different categories of the response variable, which is a measurement known as *impurity*. As the tree is being generated, it is desirable to decrease the level of impurity until ideally there is only one category at a terminal node (a node with no children).

There are three primary methods for calculating impurity: *misclassification, Gini,* and *entropy*. In the following examples the entropy calculation will be used; however, the other methods give similar results. To illustrate the use of the entropy calculation, a set of 10 observations with two possible response values ("hot" and "cold") are used (Table 5.16). All possible scenarios for splitting this set of 10 observations are shown: Scenario 1 through 11. In scenario 1, all 10 observations have value "cold" whereas in scenario 2, one observation has value "hot" and nine observations have value "cold." For each scenario, an entropy score is calculated. Cleaner splits result in lower scores. In scenario 1 and scenario 11 the split cleanly

TABLE 5.16   Entropy Scores According to Different Splitting Criteria

Scenario	Response Values		Entropy
	Hot	Cold	
Scenario 1	0	10	0
Scenario 2	1	9	0.469
Scenario 3	2	8	0.722
Scenario 4	3	7	0.881
Scenario 5	4	6	0.971
Scenario 6	5	5	1
Scenario 7	6	4	0.971
Scenario 8	7	3	0.881
Scenario 9	8	2	0.722
Scenario 10	9	1	0.469
Scenario 11	10	0	0

breaks the set into observations with only one value. The score for these scenarios is 0. In scenario 6, the observations are split evenly across the two values and this is reflected in a score of 1. In other cases, the score reflects how cleanly the two values are split.

The formula for entropy is

$$\text{Entropy}(S) = -\sum_{i=1}^{c} p_i \log_2 p_i$$

The entropy calculation is performed on a set of observations S. p_i refers to the fraction of the observations that belong to a particular value and c is the number of different possible values of the response variable. For example, for a set of 100 observations where the *Temperature* response variable had 60 observations with "hot" values and 40 with "cold" values, the p_{hot} would be 0.6 and the p_{cold} would be 0.4. When $p_i = 0$, then the value for $0 \log_2 (0) = 0$.

We illustrate this with the example shown in Figure 5.40. Values for entropy are calculated for each of the three splits:

Split a

Entropy $(N1) = -(10/20) \log_2 (10/20) - (10/20) \log_2 (10/20) = 1$
Entropy $(N2) = -(10/10) \log_2 (10/10) - (0/10) \log_2 (0/10) = 0$
Entropy $(N3) = - (0/10) \log_2 (0/10) - (10/10) \log_2 (10/10) = 0$

Split b

$$\text{Entropy (N1)} = -(10/20) \log_2 (10/20) - (10/20) \log_2 (10/20) = 1$$
$$\text{Entropy (N2)} = -(5/10) \log_2 (5/10) - (5/10) \log_2 (5/10) = 1$$
$$\text{Entropy (N3)} = -(5/10) \log_2 (5/10) - (5/10) \log_2 (5/10) = 1$$

Split c

$$\text{Entropy (N1)} = -(10/20) \log_2 (10/20) - (10/20) \log_2 (10/20) = 1$$
$$\text{Entropy (N2)} = -(1/9) \log_2 (1/9) - (8/9) \log_2 (8/9) = 0.503$$
$$\text{Entropy (N3)} = -(9/11) \log_2 (9/11) - (2/11) \log_2 (2/11) = 0.684$$

In order to determine the best split, we now need to calculate a ranking based on how cleanly each split separates the response data. This is calculated based on the impurity before and after the split. The formula for this calculation is

$$\text{Gain} = \text{Entropy(parent)} - \sum_{j=1}^{k} \frac{N(v_j)}{N} \text{Entropy}(v_j)$$

where N is the number of observations in the parent node, k is the number of possible resulting nodes, $N(v_j)$ is the number of observations for each of the j child nodes, and v_j is the set of observations for the jth node. It should be noted that the Gain formula can be used with other impurity metrics by replacing the entropy calculation.

In the example described throughout this section, the gain values are calculated and shown in Figure 5.41.

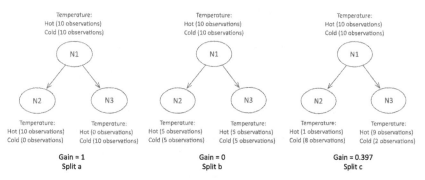

FIGURE 5.41   Calculation of gain for each split.

Gain(Splita) = $1 - (((10/20)\,0) + ((10/20)\,0)) = 1$
Gain(Splitb) = $1 - (((10/20)\,1) + ((10/20)\,1)) = 0$
Gain(Split c) = $1 - (((9/20)\,0.503) + ((11/20)\,0.684)) = 0.397$

The criterion used in **Split a** is selected as the best splitting criteria.

During the tree generation process, the method examines all possible splitting values for all splitting variables, calculates a gain function, and selects the best splitting criterion.

Scoring Splits for Continuous Response Variables   When the response variable is continuous, one popular method for ranking the splits uses the sum of the squares of error (*SSE*). The resulting split should ideally result in sets where the response values are close to the mean of the group. The lower a group's *SSE* value is, the closer that group's values are to the mean of the set. For each potential split, a *SSE* value is calculated for each resulting node. A score for the split is calculated by summing the *SSE* values of each resulting node. Once all splits for all variables are computed, then the split with the lowest score is selected.

The formula for *SSE* is

$$SSE = \sum_{i=1}^{n} (y_i - \bar{y})^2$$

For a subset of n observations, the *SSE* value is computed where y_i is the individual value for the response and $\bar{y}$ is the average value for the subset. To illustrate, the data in Table 5.17 is processed to identify the best split. The variable *Weight* is assigned as a splitting variable and *MPG* will be

TABLE 5.17   Table of Eight Observations with Values for Two Variables

Observations	Weight	MPG
A	1,835	26
B	1,773	31
C	1,613	35
D	1,834	27
E	4,615	10
F	4,732	9
G	4,955	12
H	4,741	13

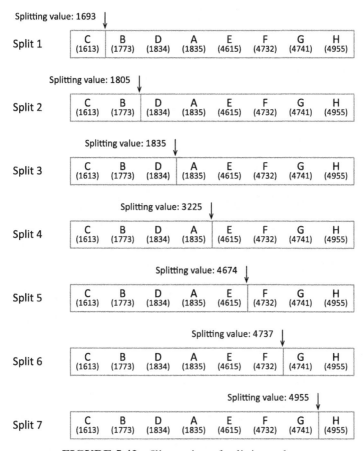

FIGURE 5.42   Illustration of splitting values.

used as the response variable. A series of values is used to split the variable *Weight*: 1,693, 1,805, 1,835, 3,225, 4,674, 4,737, and 4,955. These values are the midpoint between each pair of values (after sorting) and were selected because they divided the data set into all possible two-ways splits, as illustrated in Figure 5.42. In this example, we will only calculate a score for splits which result in three or more observations, that is, Split 3, Split 4, and Split 5. The *MPG* response variable is used to calculate the score.

Split 3

For the subset where *Weight* is less than 1835 (C, B, D):

$$\text{Average} = (35 + 31 + 27)/3 = 31$$
$$\text{SSE} = (35 - 31)^2 + (31 - 31)^2 + (27 - 31)^2 = 32$$

For the subset where *Weight* is greater than or equal to 1835 (A, E, F, H, G):

$$\text{Average} = (26 + 10 + 9 + 13 + 12)/5 = 14$$
$$\text{SSE} = (26 - 14)^2 + (10 - 14)^2 + (9 - 14)^2 + (13 - 14)^2$$
$$+ (12 - 14)^2 = 190$$

Split score $= 32 + 190 = 222$

Split 4

For the subset where *Weight* is less than 3225 (C, B, D, A):

$$\text{Average} = (35 + 31 + 27 + 26)/4 = 29.75$$
$$\text{SSE} = (35 - 29.75)^2 + (31 - 29.75)^2 + (27 - 29.75)^2$$
$$+ (26 - 29.75)^2 = 50.75$$

For the subset where *Weight* is greater than or equal to 3225 (E, F, H, G):

$$\text{Average} = (10 + 9 + 13 + 12)/4 = 11$$
$$\text{SSE} = (10 - 11)^2 + (9 - 11)^2 + (13 - 11)^2 + (12 - 11)^2 = 10$$

Split score $= 50.75 + 10 = 60.75$

Split 5

For the subset where *Weight* is less than 4674 (C, B, D, A, E):

$$\text{Average} = (35 + 31 + 27 + 26 + 10)/5 = 25.8$$
$$\text{SSE} = (35 - 25.8)^2 + (31 - 25.8)^2 + (27 - 25.8)^2 + (26 - 25.8)^2$$
$$+ (10 - 25.8)^2 = 362.8$$

For the subset where *Weight* is greater than or equal to 4674 (F, H, G):

$$\text{Average} = (9 + 13 + 12)/3 = 11.33$$
$$\text{SSE} = (9 - 11.33)^2 + (13 - 11.33)^2 + (12 - 11.33)^2 = 8.67$$

Split score $= 362.8 + 8.67 = 371.47$

In this example, Split 4 has the lowest score and would be selected as the best split.

5.4.4 Example

In the following example, a set of 392 cars is analyzed using a decision tree. Two variables were used to split nodes in the tree: *Horsepower, Weight. MPG* (miles per gallon) was used to guide the generation of the tree. A decision tree (Figure 5.43) was automatically generated using a 40 node minimum as the terminating criterion.

The leaf nodes of the tree can be interpreted using a series of rules. The decision points that are traversed in getting to the node are the rule conditions. The average *MPG* value for the leaf nodes will be interpreted here as low (less than 22), medium (22–26), and high (greater than 26). The following two example rules can be extracted from the tree:

Node A

IF *Horsepower* <106 AND
Weight < 2067.5
THEN *MPG* is high

Node B

IF *Horsepower* <106 AND
Weight 2067.5 – 2221.5
THEN *MPG* is high

In addition to grouping data sets, decision trees can also be used in making predictions and this will be reviewed in Chapter 6.

EXERCISES

Patient data was collected concerning the diagnosis of cold or flu (Table 5.18).

1. Calculate the Jaccard distance (replacing None with 0, Mild with 1, and Severe with 2) using the variables: *Fever, Headaches, General aches, Weakness, Exhaustion, Stuffy nose, Sneezing, Sore throat, Chest discomfort,* for the following pairs of patient observations:
   (a) 1326 and 398
   (b) 1326 and 1234
   (c) 6377 and 2662

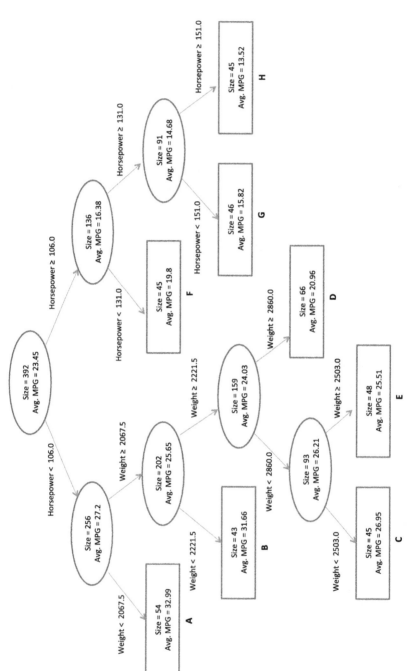

FIGURE 5.43 Decision tree generated using Horsepower and Weight as splitting values and guided by MPG.

TABLE 5.18 **Table of Patient Records**

Patient ID	Fever	Headaches	General Aches	Weakness	Exhaustion	Stuffy Nose	Sneezing	Sore Throat	Chest Discomfort	Diagnosis
1326	None	Mild	None	None	None	Mild	Severe	Severe	Mild	Cold
398	Severe	Severe	Severe	Severe	Severe	None	None	Severe	Severe	Flu
6377	Severe	Severe	Mild	Severe	Severe	Severe	None	Severe	Severe	Flu
1234	None	None	None	Mild	None	Severe	None	Mild	Mild	Cold
2662	Severe	Severe	Mild	Severe	Severe	Severe	None	Severe	Severe	Flu
9477	None	None	None	Mild	None	Severe	None	Severe	None	Cold
7286	Severe	Severe	Severe	Severe	Severe	None	None	Severe	Severe	Flu
1732	None	None	None	None	None	Severe	Severe	None	Mild	Cold
1082	None	Mild	Mild	None	None	Severe	Severe	Severe	Severe	Cold
1429	Severe	Severe	Severe	Mild	Mild	None	Severe	None	Severe	Flu
14455	None	None	None	Mild	None	Severe	Mild	Severe	None	Cold
524	Severe	Mild	Severe	Mild	Severe	None	Severe	None	Mild	Flu
1542	None	None	Mild	Mild	None	Severe	Severe	Severe	None	Cold
8775	Severe	Severe	Severe	Severe	Mild	None	Severe	Severe	Severe	Flu
1615	Mild	None	None	Mild	None	Severe	None	Severe	Mild	Cold
1132	None	None	None	None	None	Severe	Severe	Severe	Severe	Cold
4522	Severe	Mild	Severe	Mild	Mild	None	None	None	Severe	Flu

2. The patient observations described in Table 5.18 are being clustered using agglomerative hierarchical clustering. The Euclidean distance is used to calculate the distance between observations using the following variables: *Fever, Headaches, General aches, Weakness, Exhaustion, Stuffy nose, Sneezing, Sore throat, Chest discomfort* (replacing None with 0, Mild with 1, and Severe with 2). The average linkage joining rule is being used to create the hierarchical clusters. During the clustering process observations 6377 and 2662 are already grouped together. Calculate the distance from observation 398 to this group.

3. A candidate rule has been extracted using the associative rule method:

   If *Exhaustion* = None AND

   Stuffy nose = Severe

   THEN *Diagnosis* = cold

   Calculate the support, confidence, and lift for this rule.

4. Table 5.18 is to be used to build a decision tree to classify whether a patient has a cold or flu. As part of this process the Fever column is being considered as a splitting point. Two potential splitting values are being considered:

   (a) Where the data is divided into two sets when (1) Fever is none and (2) Fever is mild and severe.

   (b) Where the data is divided into two sets when (1) Fever is severe and (2) Fever is none and mild.

   Calculate the gain for each of these splits using the entropy impurity calculation.

FURTHER READING

For additional information on general data mining approaches to grouping and outlier detection, see Han et al. (2012) and Hand et al. (2001). Everitt et al. (2011) and Myatt & Johnson (2009) provide further details about similarity methods and approaches to clustering, with Fielding (2007) focusing on clustering and classification methods and their application to bioinformatics and the biological sciences. In addition, Hastie et al. (2009) covers in detail additional grouping approaches.

CHAPTER 6

BUILDING MODELS FROM DATA

6.1 OVERVIEW

In Chapter 4, we looked at different ways to understand and quantify relationships between variables. Is there a relationship between age and cholesterol levels? Do patients in a clinical trial taking a drug have improved outcomes versus patients taking a placebo? Formal ways to describe, encode, and test if and how one or more variables relate to others is to build and evaluate models from the data. These models describe important relationships in the data, including the strength and direction—positive or negative—of the relation. The models can encode linear and nonlinear relationships in the data. They can also be used to confirm a hypothesis about relationships. All these uses help to summarize and understand the data. However, one of the most widely used applications of a model is for making predictions. For example, a data set of historical purchases along with customer geographical and demographic data (such as the customer's age, location, salary, and so on) could be collected and used to generate a model that encodes what type of products clients purchase. Once the model is built, it could be used to identify from a list of potential clients those most likely

Making Sense of Data I: A Practical Guide to Exploratory Data Analysis and Data Mining,
Second Edition. Glenn J. Myatt and Wayne P. Johnson.
© 2014 John Wiley & Sons, Inc. Published 2014 by John Wiley & Sons, Inc.

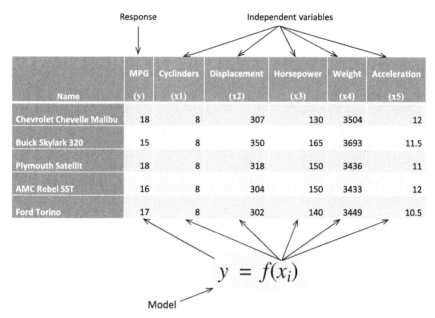

FIGURE 6.1   Illustration of response versus independent variables.

to make a purchase, and customers on this prioritized list could be targeted with marketing material or other promotions.

In this chapter, we will review how models can be built from data sets. A model is usually built to predict values for a specific variable. For example, were a data set composed of historical data containing attributes of pharmaceuticals and their observed side effects to be collected, a model can be generated from this data to predict the side effects from the pharmaceuticals' attributes.

A variable that a model is to predict is often referred to as a y-variable or *response* variable. The variables that will be encoded in the model and used in predicting this response are referred to as the x-variables or the *independent variables*. In Figure 6.1, a data table composed of cars is used to generate a model. Because we want the model to predict the car's fuel efficiency, we have chosen the response variable to be miles per gallon (*MPG*). Other variables will be used as independent variables (x-variables). In this case, these will be *Cylinders* (x_1), *Displacement* (x_2), *Horsepower* (x_3), *Weight* (x_4), and *Acceleration* (x_5). A generalized format for the model is shown where some function of the independent variables (x_i) is used to predict the response (y), which in this case is *MPG*.

Models built to predict categorical variables (such as a binary variable or a nominal variable) are referred to as *classification* models, whereas models that predict continuous variables are called *regression* models. There are many ways to generate classification and regression models. For example, a *classification tree* is a method for building a classification model while a *multiple linear regression* is a method for building a *regression model.* Specific approaches may have restrictions relating to the types of variables that can be used in the model as, for example, a model that requires continuous variables to have a normal frequency distribution. For certain types of models it is possible to fine-tune the performance of the model by varying different parameters. In building a model, it will be important to understand the restrictions placed on the types of independent or response variables or both, as well as how to optimize the performance of the model by varying the values of the parameters. Another way in which approaches to modeling differ is in the ease of access to the internal calculations, otherwise known as the *transparency* of the model, in order to explain the results: is it possible to understand how the model calculated a prediction or is the model a "black box" that only calculates a prediction result with no corresponding explanation? Issues related to transparency may be important in explaining the results when the model is deployed in certain situations.

Although the response variable is known, when building models it is not always apparent beforehand which variables should be used as independent variables. Therefore, the selection of the independent variables is an important step in building a model. A good model will make reliable predictions, be plausible, and use as few independent variables as possible. In Chapter 4, we reviewed different visualizations and metrics to use in understanding relationships in the data, such as scatterplots, contingency tables, *t*-tests, Chi-Square tests, and so on. These approaches can be used to prioritize candidate independent variables to use in building a model, especially where there are many variables to consider. However, care should be taken when using statistical tests to prioritize large numbers of potential independent variables as a correction may need to be used (see the Further Reading section of this chapter for more information). For example, a matrix of scatterplots could be used to visually identify which variables have the strongest relationship to the response variable. In addition, knowledge of the problem can also guide the choice of variables to use in the models. Alternatively, we could build multiple models with different combinations of independent variables and select the best fitting model.

Another issue to consider when selecting independent variables is the relationship between the independent variables. Combinations of variables that have strong relationships to each other should be avoided since they will be essentially encoding the same relationship to the response. Including all the variables from each group of strongly related variables produces overly complex models (violating the "as simple as possible rule") and with some approaches to modeling can produce results that are difficult to interpret.

In developing a model, it may also be necessary to use derived variables, that is, a new variable that is a function of one or more variables. For example, if the model expects the variables of a data set to have a normal frequency distribution and some variables have an exponential frequency distribution, it may be necessary to create new variables using a log transformation. As another example, because most modeling methods require numeric data, if a data set has nominal variables that will be used in the model, the values of these variables must be transformed into numbers. For example, if *color* is an important variable with values "Blue," "Green," "Red," and "Yellow," *color* could be transformed into a series of binary dummy variables as described in Chapter 3.

In this chapter, we discuss how to generate models from data sets. The data set used to build a model is referred to as the *training set*. To objectively test the performance of a generated model, a *test* set with observations different from those in the training set is used to test how well the model performs. The model uses the values of each observation in the test set to predict a value for the response variable. From these predictions, a variety of metrics, such as the number of correct versus incorrect predictions made, are used to assess the accuracy of the model. The use of training and test sets is illustrated in Figure 6.2.

A good way to build and test a model would be to use all the observations in the original data set as the training set to build the model and to use new, independent observations as the test set to measure accuracy. However, because the number of available test sets is often small, a common way to test the performance of a model is to use a category of methods called *cross-validation*. In the *k-fold partitioning* method, the original data set is divided

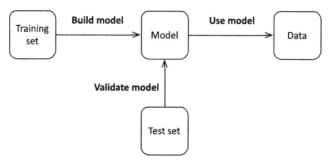

FIGURE 6.2   Use of training sets to build models and test sets to assess their performance.

into k equally sized partitions. The model is measured k times. In the first iteration, one of the partitions is selected as the test set and the remaining partitions comprise the training set. The model is tested and an accuracy score is generated. In each subsequent iteration, a partition different from any already used as a test set is selected as the test set and the remaining partitions become the training set. Another score is calculated. At the end of this process, the accuracy of the model is based on the average of the k scores. For example, suppose we partition a data set into 10 partitions where each partition consists of observations randomly selected from the data set. In each of the 10 iterations, we designate one partition (10% of the data set) as the test set and the other 9 partitions (90% of the data set) as the training set. At the end of the 10 iterations, an average of the 10 scores is used to assess the model's accuracy. Taking k-fold partitioning to an extreme would result in the case where k is the number of observations in the data set and each partition contains a single observation. This is a cross-validation method known as *leave-one-out*.

In cross-validation, each partition will have been used as a test set or, in other words, every observation in the data set will have been tested once. This ensures that a prediction will be calculated for every observation in the data set and avoids introducing *bias* into a model. Bias is a measure of the model's accuracy and indicates how close the predictions of the response value made by the model are to the actual response value of new observations. It can be introduced when models become overly complex by optimizing the model for just the training set used to build the model. When the performance is tested for these overtrained models against either a separate test set or through cross-validation, the performance will be poorer. In cross-validation methods, bias can be introduced when training sets overlap (some observations are used more than once) or the combined training sets do not cover the data set (some observations are never used).

For classification models, one way to assess the performance of a model is to look at the results of applying the models (such as the results from a test set or the cross-validation results) and determine how many observations are correctly or incorrectly classified. The accuracy or *concordance* of the model is based on the proportion or percentage of correctly predicted observations in comparison to the whole set. For example, if the test set contained 100 observations and the model predicted 78 correctly (22 incorrectly), then the concordance would be 78/100 or 78%.

A common type of classification model is a model to predict a binary response, where a true response is coded as 1 and a false response is coded as 0. For example, a model could be built to predict, based on geological data, whether there is evidence of an oil deposit, with a true response

TABLE 6.1 Contingency Table Summarizing the Correct and Incorrect Predictions from a Binary Classification Model

		Actual		
		True (1)	False (0)	
Prediction	True (1)	True positives (TP)	False positive (FP)	*Number of observations predicted true* (1)
	False (0)	False negatives (FN)	True negatives (TN)	*Number of observations predicted false* (0)
		Number of actual true (1) *values*	*Number of actual false* (0) *values*	*Total observations*

encoded as a value of 1 when there is evidence for an oil deposit and a false response as a value of 0 if there is not. A good way to evaluate a classification model's performance is through a contingency table summarizing the number of correct and incorrect classifications. The number of correctly predicted positive observations (*true positives* or TP) and the number of correctly predicted negative observations (*true negatives* or TN) is shown in Table 6.1. In addition, the number of positive predictions that are incorrect is referred to as *false positives* (or FP) and the number of negative predictions that are incorrect is referred to as *false negatives* (or FN) are also summarized in Table 6.1.

To illustrate the difference in how well models are able to predict positive and negative values, the results from three models are presented in Figure 6.3. Model 1 correctly predicts 75% (36 out of 48 positives and 39 out of 52 negatives). In Model 1, the number of false positives (12) and the number of false negatives (13) are quite similar. Model 2 has an overall concordance of 80% with few false negatives (only 3) but a larger number of false positives (17). In Model 3, the balance of the false positives and false negatives is more biased toward false negatives (22) than to false positives (9). However, the overall concordance values do not reflect biases in the model's ability to predict true or false values. To better assess the overall performance of a binary classification model, it is necessary to calculate additional metrics. Two commonly used calculations are *sensitivity* and *specificity*:

$$\text{Sensitivity} = \text{TP}/(\text{TP} + \text{FN})$$
$$\text{Specificity} = \text{TN}/(\text{TN} + \text{FP})$$

Model 1

Prediction	Actual 1	Actual 0	Total
1	36	12	48
0	13	39	52
Total	49	51	100

Concordance	75.0%
Sensitivity	73.5%
Specificity	76.5%

Model 2

Prediction	Actual 1	Actual 0	Total
1	31	17	48
0	3	49	52
Total	34	66	100

Concordance	80.0%
Sensitivity	91.2%
Specificity	74.2%

Model 3

Prediction	Actual 1	Actual 0	Total
1	40	9	49
0	22	39	61
Total	62	48	110

Concordance	71.8%
Sensitivity	64.5%
Specificity	81.3%

FIGURE 6.3 Contingency tables and performance metrics for three models.

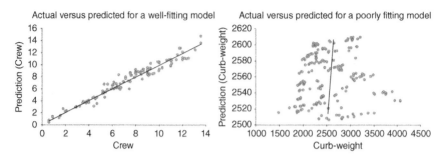

FIGURE 6.4   Scatterplot for a well- and poorly fitting regression model.

Sensitivity generally describes how well a model predicts positives, whereas specificity generally describes how well a model predicts negatives. The values for sensitivity and specificity in Model 1 are similar since the number of false positives and false negatives are similar. In Model 2, the sensitivity value is high (91%) which reflects the low number of false negatives, whereas the specificity in Model 3 is high reflecting the lower number of false positives.

In assessing the performance of a regression model, a scatterplot with axes showing the actual values and the predicted values is a useful way to start to understand the performance of the model. Models that accurately predict a response variable have points close to and evenly distributed about a straight line, as shown in the left scatterplot in Figure 6.4, whereas poor performing models have points scattered as illustrated in the right scatterplot in Figure 6.4. Scatterplots can also help to understand if there are observations that will be poorly predicted for a given model. These observations appear as points that do not fall close to the best fit line. If the scatterplot trend has a nonlinear shape, then the model is not capturing the nonlinear relationships and one or more of the variables included in the model may require a data transformation or a nonlinear modeling approach selected. The *error* or *residual* is the difference between the predicted value and the actual value. An overall score based on these residual values can be helpful in assessing the relative performance of different models. Since a residual can contain positive and negative values, it is usual to calculate an overall assessment of the residuals, such as the sum of the absolute residual or the square of the residual.

In using a model in practice, it is not advisable to apply a model to data sets that are not similar to those used in building the model (*extrapolation*). It is usual to place some restriction on the variables of the data sets that will be input to the model, such as the requirement that values not be outside the range of the training set variables.

The following sections outline a number of common and diverse approaches to building models: *linear regression*, *logistic regression*, *k-nearest neighbors*, and *classification and regression trees* (*CART*). The chapter ends with a review of other approaches to building models and additional information on resources for these topics.

6.2 LINEAR REGRESSION

6.2.1 Overview

The following section discusses how to generate *linear* models to describe a relationship between one or more independent variables and a single response variable. For example, we could build a linear regression model to predict cholesterol levels using data about a patient's age. This model will likely be a poor predictor of cholesterol levels; however, incorporating more information, such as body mass index (BMI) may result in a model that provides a better prediction of cholesterol levels. Using a single independent variable is referred to as *simple linear regression*, whereas using more than one independent variable is referred to as *multiple linear regression*. Although these models do not make causal inferences, they are useful for understanding how a set of independent variables is associated with a response variable. The following sections describe how to generate and assess linear regression models and test the assumptions about the model.

6.2.2 Fitting a Simple Linear Regression Model

A simple linear regression model can be generated where there is a linear relationship between two variables. For example, Figure 6.5 shows the relationship between the independent variable *Age* and the response variable *Blood fat content*. The diagram shows a high degree of correlation between the two variables. As variable *Age* increases, response variable *Blood fat content* increases proportionally. A straight line, representing a linear model, can be drawn through the center of the points.

This straight line can be described using the formula

$$y = b_0 + b_1 x$$

where b_0 is the point of intersection with the y-axis and b_1 is the slope of the line, which is shown graphically in Figure 6.6. The simple linear regression model is usually shown with an error term; however, it is not included here to simplify the example.

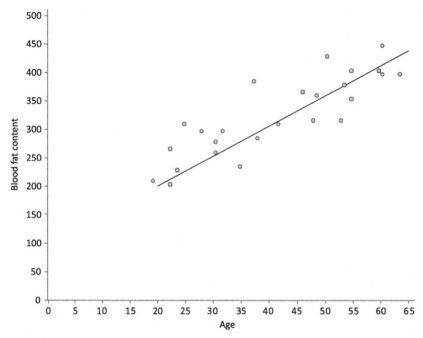

FIGURE 6.5   A straight line drawn through the relationship between variables *Age* and *Blood fat content*.

The point at which the line intercepts with the *y*-axis is noted (approximately 100) and the slope of the line is calculated (approximately 5.3). For this data set, an approximate formula for the relationship between *Age* and *Blood fat content* is

$$Blood\ fat\ content = 100 + 5.3 \times Age$$

Parameters b_0 and b_1 can be derived manually by drawing a line through the points in the scatterplot and then visually inspecting where the line crosses the *y*-axis (b_0) and measuring the slope (b_1), as previously described. The *least-squares method* is able to calculate these parameters automatically. The formula for calculating a slope (b_1) is

$$b_1 = \frac{\sum_{i=1}^{n}(x_i - \bar{x})(y_i - \bar{y})}{\sum_{i=1}^{n}(x_i - \bar{x})^2}$$

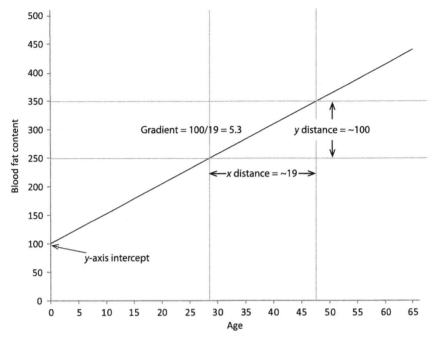

FIGURE 6.6 Deriving the straight line formula from the graph.

where x_i and y_i are the individual values for the independent variable (x) and the response variable (y), and where $\bar{x}$ is the mean of x and $\bar{y}$ is the mean of y.

The formula for calculating the intercept with the y-axis is

$$b_0 = \bar{y} - b_1 \bar{x}$$

The slope and intercept are calculated using the data from Table 6.2. The mean of x is 39.12 and the mean of y is 310.72.

$$Slope\,(b_1) = 19{,}157.84/3{,}600.64$$
$$Slope\,(b_1) = 5.32$$

$$Intercept\,(b_0) = 310.72 - (5.32 \times 39.12)$$
$$Intercept\,(b_0) = 102.6$$

Hence the equation is

$$Blood\ fat\ content = 102.6 + 5.32 \times Age$$

TABLE 6.2 Calculation of Linear Regression with Least Square Method

X	Y	$(x_i - \bar{x})$	$(y_i - \bar{y})$	$(x_i - \bar{x})(y_i - \bar{y})$	$(x_i - \bar{x})^2$
46	354	6.88	43.28	297.7664	47.3344
20	190	−19.12	−120.72	2,308.1664	365.5744
52	405	12.88	94.28	1,214.3264	165.8944
30	263	−9.12	−47.72	435.2064	83.1744
57	451	17.88	140.28	2,508.2064	319.6944
25	302	−14.12	−8.72	123.1264	199.3744
28	288	−11.12	−22.72	252.6464	123.6544
36	385	−3.12	74.28	−231.7536	9.7344
57	402	17.88	91.28	1,632.0864	319.6944
44	365	4.88	54.28	264.8864	23.8144
24	209	−15.12	−101.72	1,538.0064	228.6144
31	290	−8.12	−20.72	168.2464	65.9344
52	346	12.88	35.28	454.4064	165.8944
23	254	−16.12	−56.72	914.3264	259.8544
60	395	20.88	84.28	1,759.7664	435.9744
48	434	8.88	123.28	1,094.7264	78.8544
34	220	−5.12	−90.72	464.4864	26.2144
51	374	11.88	63.28	751.7664	141.1344
50	308	10.88	−2.72	−29.5936	118.3744
34	220	−5.12	−90.72	464.4864	26.2144
46	311	6.88	0.28	1.9264	47.3344
23	181	−16.12	−129.72	2,091.0864	259.8544
37	274	−2.12	−36.72	77.8464	4.4944
40	303	0.88	−7.72	−6.7936	0.7744
30	244	−9.12	−66.72	608.4864	83.1744
			Sum	19,157.84	3,600.64

These coefficient values are close to the values calculated using the manual approach.

For each value of the x-variable, the corresponding y-variable value (taken from the straight line) represents the expected mean y value. The actual values will fall above and below the straight line since the line represents the mean.

Once a formula for the straight line has been established, predicting values for the y response variable based on the x independent variable can be easily calculated. However, the formula should only be used for values of the x variable within the range in which the formula was derived. In this example, *Age* values should only be between 20 and 60. A prediction for *Blood fat content* based on *Age* can be calculated, for example, for an individual whose age is 33, the *Blood fat content* would be predicted as

$$Blood\ fat\ content = 102.6 + 5.32 \times 33 = 278.16$$

The slope can be interpreted as the average amount that the *Blood fat content* changes for each unit (year) change in *Age*. The intercept represents the average value of the *Blood fat content* when *Age* is zero; however, in this case an *Age* of zero is out of the range of meaningful interpretation.

6.2.3 Fitting a Multiple Linear Regression Model

In most practical situations, a simple linear regression is not sufficient because the models will need more than one independent variable. The general form for a multiple linear regression equation is a linear function of the independent variables:

$$y = b_0 + b_1 x_{1i} + b_2 x_{2i} + \cdots + b_k x_{pi} + e_i$$

where the response variable (y) is shown with p independent variables (x-variables), b_0 is a constant value, k is the number of coefficients of the independent variables, and e_i refers to an error term measuring the unexplained variation or noise in the linear relationship.

The set of coefficients are calculated as part of the model building process to minimize the overall differences between the observed and the predicted response values. Since the mathematics for computing all but the simplest models make it impossible to compute by hand, software tools are typically used to perform the computation. This results in an equation where the coefficients are estimated:

$$\hat{y} = \hat{b}_0 + \hat{b}_1 x_1 + \hat{b}_2 x_2 + \cdots + \hat{b}_p x_k$$

In this equation, the coefficients are shown with a "hat" to represent that they are estimated. This form was not presented for the simple linear regression example to simplify the example. For example, a data set of cruise ships is used to build a model to predict the number of crew required in hundreds (*Crew*), with the first five rows (out of a total of 154) shown in Table 6.3 from Winner (2013). A multiple linear regression model is built using the variable *Cabins* (number of cabin on the ship in hundreds) and *Passenger density* (the passenger to space ratio). The model equation is

$$Crew = -0.423 + 0.75 \times Cabins + 0.0377 \times Passenger\ density$$

TABLE 6.3   Sample from the Data Set of Cruise Ships

Order	Ship Name	Cruise Line	Cabins	Passenger Density	Crew
1	Journey	Azamara	3.55	42.64	3.55
2	Quest	Azamara	3.55	42.64	3.55
3	Celebration	Carnival	7.43	31.8	6.7
4	Destiny	Carnival	13.21	38.36	10
5	Ecstasy	Carnival	10.2	34.29	9.2

It is now possible to estimate the number of crew that would be required for a cruise ship. For example, a cruise ship with 11.1 *cabins* (1,100 actual cabins) and *passenger* density of 42.7 would require a *crew* of 9.51 or 951 (since the *Crew* variable is based on hundreds):

$$Crew = -0.423 + 0.75 \times 11.1 + 0.0377 \times 42.7 = 9.51$$

The individual coefficients, similar to the simple linear regression situation, can be interpreted as slopes of the independent variables. In assessing the slope (coefficient) for a particular variable, the slope represents the average amount of increase (for positive slope values) or decrease (for negative slope values) of the response per one unit increase/decrease in the variable under consideration (keeping the other variables constant). For example, if *Passenger density* is held constant then an increase in the *Cabins* variable of 1.0 would mean an increase in the *Crew* variable of 0.75 (the coefficient or slope value for *Cabins*).

6.2.4 Assessing the Model Fit

As part of the process of generating the linear regression model, or in other words estimating the model coefficients, a set of statistics are usually generated that help to understand the overall accuracy of the model. The *residual* ($\hat{e}$) is an error term representing the difference between the observed value (y) and the predicted value ($\hat{y}$):

$$\hat{e} = y - \hat{y}$$

For example, Table 6.4 shows Table 6.3 with two additional columns. Column ($\hat{y}$) was added to show the predicted value for each row. Since the actual and predicted values differ, another column was added to show the residual ($\hat{e}$). For the ship "Journey," the actual value for *Crew* is 3.55, whereas the predicted value is slightly higher (3.85) resulting in a negative residual (−0.3). For the ship "Celebration," the actual value is 6.7, whereas the predicted value is 6.35 and so the residual is positive (0.35).

TABLE 6.4   Cruise Ship Data Annotated with Predicted Values and Calculated Residuals

Order	Ship Name	Cruise Line	Cabins	Passenger Density	Crew	Predicted $(\hat{y})$	Residual $(\hat{e})$
1	Journey	Azamara	3.55	42.64	3.55	3.85	−0.3
2	Quest	Azamara	3.55	42.64	3.55	3.85	−0.3
3	Celebration	Carnival	7.43	31.8	6.7	6.35	0.35
4	Destiny	Carnival	13.21	38.36	10	10.9	−0.9
5	Ecstasy	Carnival	10.2	34.29	9.2	8.52	0.68

Looking at the residual—the difference between the prediction and the actual value—helps to better understand how well the model is performing.

The *sum of squares total* (*SST*) is a measure of the variation of the y-values about their mean:

$$SST = \sum_{i=1}^{n} (y_i - \bar{y})^2$$

Of this total variation, part of the variation is explained and attributable to the relationship between the x-variables and the y-variable or *sum of squares due to regression* (*SSR*). This formula looks at the differences between the predicted values (calculated from the regression equation) and the average y-values:

$$SSR = \sum_{i=1}^{n} (\hat{y}_i - \bar{y})^2$$

The other part of the total variation is unexplained by the model and hence attributable to the error or *sum of squares* of *error* (*SSE*) and looks at the differences between the actual y-values (y_i) and the predicted y-values ($\hat{y}_i$):

$$SSE = \sum_{i=1}^{n} (y_i - \hat{y}_i)^2$$

Since the total sum of squares (*SST*) is composed of the explained (*SSR*) and the unexplained (*SSE*) variations, it follows that the value of *SST* can be derived from *SSR* and *SSE*:

$$SST = SSR + SSE$$

In the example of the cruise ship linear model *SSR* is 1,484, SSE is 88.3 and *SST* is 1,573. The *coefficient of determination* (R^2) represents the proportion of the variation that is explained by the set of x-variables in the model and is the ratio of SSR to SST:

$$R^2 = \frac{SSR}{SST}$$

Using the values from the cruise ship example, R^2 would be

$$R^2 = \frac{1{,}484}{1{,}573} = 0.94$$

94% of the variable *Crew* can be explained by the variability in the independent variables (*Cabins* and *Passenger density*) with 5.7% attributable to something else. R^2 values vary between 0 and 1. The closer the values are to 1, the more accurate are the predictions of the model; we say these models have a *closer fit*. With multiple linear regression, an *adjusted R^2* value (R^2_{adj}) is usually considered to better account for the multiple independent variables used in the analysis as well as the sample size. Its formula is

$$R^2_{adj} = 1 - \left[\frac{(1 - R^2)(n - 1)}{n - k - 1} \right]$$

where n is the number of observations and k is the number of independent variables. In the cruise ship example, the value of R^2_{adj} is

$$R^2_{adj} = 1 - \left[\frac{(1 - 0.94)(154 - 1)}{154 - 2 - 1} \right] = 0.94$$

It is also a typical practice to calculate the *standard error of the estimate* ($s_{y.x}$), which is a measure of the variation of the y-values about the regression line. This value is interpreted in a similar manner to standard deviation and has the formula

$$s_{y.x} = \sqrt{\frac{SSE}{n - 2}}$$

In the cruise ship example, s would be

$$s_{y.x} = \sqrt{\frac{88.3}{154 - 2}} = 0.76$$

The value indicates the model's accuracy: the larger the value for the standard error of estimate, the lower the precision.

As long as the linear regression assumptions (described in Section 6.2.5) are not seriously violated, inferences can be made. A t-test is used to determine whether there is a significant linear relationship between a specific independent variable and the response. As described earlier, the null and alternative hypothesis should be defined where the null hypothesis is that there is no linear relationship and the alternative hypothesis states that there is. If the null hypothesis can be rejected, then there is evidence of a linear relationship. The following formula for calculating the t-value is used:

$$t = \frac{\hat{b}_i - b_i}{s_{b_i}}$$

where

$$s_{b_i} = \frac{s_{x.y}}{\sqrt{\sum(x_i - \bar{x})^2}}$$

For the *Cabins* variable, the t-value is calculated as follows using 0 for b_i to represent that there is no relationship:

$$t = \frac{0.75 - 0}{0.0151} = 49.7$$

For the *Passenger density* variable, the t-value is calculated as

$$t = \frac{0.0377 - 0}{0.00734} = 5.14$$

Using a value for $\propto$ of 0.05, the critical t-value is ± 1.96. Since the t-values for both *Cabins* and *Passenger density* are greater than the critical value, we reject the null hypothesis and conclude that each of the two individual coefficients exhibit a significant relationship with the response variable. It is usual to calculate a p-value, which would be less than 0.0001 for both the *Cabins* and the *Passenger density* coefficient.

The F-test is used to test whether there is a statistically significant relationship between the x-variables and the y-response. Again, a null hypothesis (there is no linear relationship) and an alternative hypothesis (a linear relationship exists between at least one of the independent variables) is stated. The *F*-test makes use of the formulas for mean square regression (*MSR*) and mean square error (*MSE*):

$$MSR = \frac{SSR}{k}$$

$$MSE = \frac{SSE}{n - k - 1}$$

$$F = \frac{MSR}{MSE}$$

where *n* is the size of the data set and *k* is the degrees of freedom (as discussed in Section 4.3.5). The null hypothesis is rejected where the *F*-value is greater than the critical value of *F* based on *k* degrees of freedom (regression) and $n - k - 1$ degrees of freedom (error).

In the cruise ship example,

$$MSR = \frac{1{,}484}{2} = 742$$

$$MSE = \frac{88.3}{154 - 2 - 1} = 0.585$$

$$F = \frac{742}{0.585} = 1{,}268$$

Since this value is greater than the critical value of *F* (identified from a standard *F*-distribution table), we reject the null hypothesis and conclude that there is at least one significant relationship.

The full results of a linear regression are often presented as shown in Figure 6.7.

6.2.5 Testing Assumptions

Linear regression models are based on a series of assumptions. If a data set does not conform to these assumptions then either the model needs to be adjusted—such as applying a mathematical transformation to the data—or multiple linear regression may not be suitable for modeling the data set.

The first assumption is that of *linearity*: the relationship between the independent variables and the response variable should be linear. A scatterplot displaying the actual response values plotted against the predicted values is one approach to checking this assumption. The points on the scatterplot should be evenly distributed on both sides of the regression line. Another approach is to look at the residual values plotted against the

Coefficients					
	Coefficients	Std error	t-stat	p-value	
Intercept	−0.423	0.357	−1.18	0.2	
Cabins	0.75	0.0151	49.7	<0.0001	
Passenger density	0.0377	0.00734	5.14	<0.0001	
ANOVA					
	df	SS	MS	F	Sig
Regression	2	1484	742	1268	<0.0001
Residual	151	88.3	0.585		
Total	153	1573			
Regression Analysis					
Nos. observations	154				
R^2	0.94				
R^2_{adj}	0.94				
Standard error	0.765				

FIGURE 6.7   Summary of the multiple linear regression results.

predicted values. There should be no discernable trend in the data. Figure 6.8 is a scatterplot of the residual values plotted on the y-axis and the predicted values plotted on the x-axis showing no readily observable trend. If a nonlinear trend is observed, a mathematical transformation, such as a log transformation or the introduction of an additional x^2 value (to obtain an equation in the form $y = b_0 + b_1 x + b_2 x^2$), should be considered.

The second assumption is the *normality of the error distribution*. The error about the line of regression should be approximately normally distributed for each value of x. This assumption can be tested using either a frequency histogram, statistical measures of skewness/kurtosis, or a normal probability plot. If the residuals frequency distribution does not approximate a normal distribution, then transformations of the variables should be considered to map them more closely onto a normal distribution. The presence of outliers in the data may also affect the distribution and these

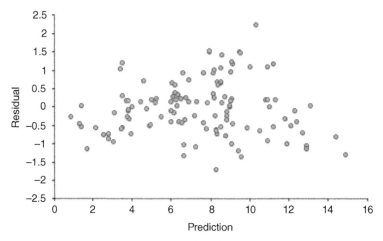

FIGURE 6.8   Scatterplot of the residuals against the predicted values.

should be checked in case there are errors or anomalies that would warrant removing them from the data set.

A third assumption is *homoscedasticity of the errors*. The variation of the error or residual across each of the independent variables should remain constant either as a function of time (in time-series data sets) or a function of the predicted value. For example, the errors in models generated from stock market data could be affected by seasonal changes or by an increase in the rate of inflation over time. There should be no discerning trend when the residuals are plotted on the *y*-axis against (1) the order in which the values were measured, (2) the predicted values, and (3) the independent variables. Trends such as the variation in the error getting larger or smaller as the values along the *x*-axis increase or decrease would suggest a violation of the homoscedasticity assumption.

The final assumption is the *independence of errors*. There should be no trend in the residuals based on the order in which the observations were collected. Again, this can be tested using a scatterplot of the residual values versus the order in which they were collected. There should be no discernable trend in the data—the observations should spread out evenly. Methods examining this assumption in more detail include the use of the Durbin–Watson statistic.

6.2.6 Selecting and Assessing Independent Variables

An important part of generating linear regression models is the selection of independent variables. It is important to start with a plausible combination of variables, which is a set that a domain expert would identify as having a relationship to the response. It is also important to generate the simplest possible model that contains only those independent variables considered necessary. A rule of thumb is to keep the number of independent variables to a relatively small set and to include at least 10 observations in the training set for every independent variable included in the model.

As described in Chapter 3, it may also be necessary to perform transformations on the pool of potential independent variables. Several techniques are available for doing this. Dummy variables can ge generated where there are nominal variables that need to be included. Continuous variables that need to be transformed into a categorical variable can use a function that incorporates a series of cutoffs identified by one or more points at which the response changes dramatically. If the relationship between a potential independent variable and the response variable needs to be converted from nonlinear to linear, it can be done by the application of transformations such as a log or exponential function. Finally, it may be necessary to introduce a new independent variable that is a function of two or more variables, such as a multiplication or a ratio.

Prior to building a prediction model, it is helpful to use exploratory data analysis methods to inspect the relationships between the variables. This includes the relationship between each independent variable under consideration and the response. It is also important to understand the relationships between each pair of independent variables because including variables strongly related to each other adds little new information to the model and makes the final model difficult to interpret.

Multiple combinations of different independent variables can be used to build a set of models from which the best performing, most plausible, and simplest model is selected. In addition, there are ways of automatically selecting the "best" combinations of independent variables (discussed in the Further Reading section) using methods such as stepwise regression.

Once the equation parameters have been selected and a model has been built, tools that generate linear regression models produce a series of statistics about the coefficients. In addition to the value of the constants used in the equations, the *standard error*, *t-stat*, and *p-value* is calculated, as discussed earlier, which can be used to help in the selection of the independent variables.

6.3 LOGISTIC REGRESSION

6.3.1 Overview

As discussed in the previous section, the multiple linear regression approach can only be used to make predictions when the response variable is continuous. It cannot be used when the response variable is categorical. Logistic regression is a popular approach to building models where the response variable is usually binary (dichotomous). For example, the response variable could indicate whether a consumer purchases a product (1 if they purchase and 0 if they do not) or whether a candidate drug is potent (1 if the candidate drug is potent and 0 if it is not). Logistic regression provides a flexible and easy-to-interpret method for building models from binary data. The following section outlines how to build, use, and assess logistic regression models.

6.3.2 Fitting a Simple Logistic Regression Model

A data set related to the presence of gold deposits (five rows of which are shown in Table 6.5) will be used to illustrate how a logistic regression model operates from Sahoo and Pandalai (1999). The data set includes observations showing measured Sb levels (log transformed) and whether there is a gold deposit within 0.5 km (1 indicates there is a gold deposit and 0 indicates there is none). The average *log(Sb level)* where *Gold deposit proximity* is 1 is 0.445 and the average value when *Gold deposit proximity* is 0 is −0.444 indicating that there is a difference between the two values; however, it does not describe the type of relationship well. To better understand this relationship, we will make the values of the *log(Sb level)* discrete and plot these values, as shown in Table 6.6 and Figure 6.9.

The relationship between the average value of the *Gold deposit proximity* variable and the *log(Sb level)* can be seen in Figure 6.10. As the values for *log(Sb level)* increase, the mean values for the *Gold deposit proximity* also rise; however, the relationship is not linear. It follows an S-shaped curve, starting at a mean value of 0 (all values are 0) and ending at a mean value of 1 (all values are 1), with a more rapid transition from the low to high mean values toward the center of the graph. The graph can never go below 0 or above 1 and the mean values at the low end of the log(Sb level) range as well as at the high log(Sb level) range are flat.

TABLE 6.5 Five Rows from the Gold Data Set

Log(Sb level)	Gold Deposit Proximity
−0.251811973	0
−0.22184875	0
−0.15490196	1
−0.096910013	0
−0.040958608	0

TABLE 6.6   The Mean Value for the Variable Gold Deposit Proximity for Different Ranges of Log(Sb Level)

Log(Sb Level) Ranges	Mean Gold Deposit Proximity
−1.1 to −0.7	0
−0.7 to −0.3	0.04
−0.3 to 0.1	0.43
0.1 to 0.5	0.89
0.5 to 0.9	0.9
0.9 to 1.3	1

To generate a model that describes the relationship between the independent variable or *x*-variable (*log(Sb level)*) and the response or *y*-variable (*Gold deposit proximity*), we will need to understand the relationship between the mean or expected value (*E*) of the response (*Y*) given a specific value for the *x* variable (*E(Y|x)*). In Figure 6.10 we are showing the mean response value on the *y*-axis. In order to map the *x* values onto the mean *y*-values we need to model this S-shaped curve which can be accomplished through a logistic formula:

$$E(Y|x) = \frac{e^{\beta_0 + \beta_1 x}}{1 + e^{\beta_0 + \beta_i x}}$$

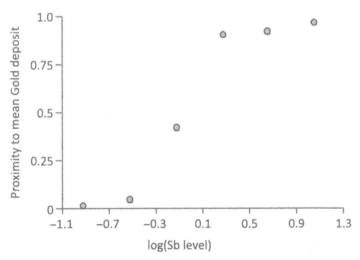

FIGURE 6.9   Graph showing how the mean values for variable Gold deposit proximity increase as values for the log-transformed Sb level increase.

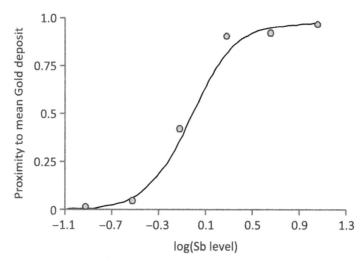

FIGURE 6.10   Shape of the graph showing how the mean values for variable Gold deposit proximity increase as values for the log-transformed Sb level increase.

In this equation, the expected value of y given x ($E(Y|x)$) is calculated where e is the exponential function and β_0, β_1 are constant values. This formula will calculate values for the $E(Y|x)$ along the "S"-shaped curve. It also ensures that values do not exceed 1 or go below 0, as shown in Figure 6.10.

The error for logistic regression has different characteristics to the error discussed in the section on linear regression. Since the values of Y can be only 1 or 0, the error is either $1 - E(Y|x)$ or $0 - E(Y|x)$, hence it follows a binomial distribution (rather than an approximate normal distribution as in the case of linear regression).

For a given data set, the beta coefficients are estimated using a *maximum likelihood method* (see Hosmer et al. (2013) for details). This process is invariably performed using computer software. For the data set illustrated in Table 6.5, the following formula is generated:

$$E(\text{Gold deposit proximity}|log(Sb\ level)) = \frac{e^{-0.0728+5.82\times log(Sb\ level)}}{1 + e^{-0.0728+5.85\times log(Sb\ level)}}$$

To calculate a value for the expected value (mean) for the *Gold deposit proximity* ($E(\text{Gold deposit proximity}|log(Sb\ level))$) we can substitute the original value with its log-transformed value in the formula.

If *log(Sb level)* was 0.4, then

$$E(Gold\ deposit\ proximity|log(Sb\ level) = 0.4) = \frac{e^{-0.0728+5.82\times0.4}}{1 + e^{-0.0728+5.85\times0.4}}$$
$$= 0.905$$

Since the expected value is close to 1, we could conclude that it is likely there will be a gold deposit within 0.5 km.

6.3.3 Fitting and Interpreting Multiple Logistic Regression Models

In almost all practical situations, multiple independent variables will be used to build a logistic regression model. The formula can be extended to accommodate p independent variables:

$$E(Y|x) = \frac{e^{\beta_0+\beta_1 x_1+\beta_2 x_2+\cdots+\beta_p x_p}}{1 + e^{\beta_0+\beta_1 x_1+\beta_2 x_2+\cdots+\beta_p x_p}}$$

For example, two measurements, As levels and Sb levels, could be collected and used to predict whether a gold deposit will be identified within 0.5 km (*Gold deposit proximity*). Again the log transformation of the original As levels and Sb level measurements will be used: *log(As levels)*, *log(Sb levels)*. By loading the data set into a computational package, it is possible to derive a formula to predict the expected value for *Gold deposit proximity*:

$$E(Y|x) = \frac{e^{-1.84+5.2\log(As\ levels)+3.5\log(Sb\ levels)}}{1 + e^{-1.84+5.2\log(As\ levels)+3.5\log(Sb\ levels)}}$$

The predicted value based on actual As levels and Sb levels would be

$$E(Y|x) = \frac{e^{-1.84+5.2\times0.1+3.5\times0.05}}{1 + e^{-1.84+5.2\times0.1+3.5\times0.05}} = 0.24$$

where a *log(As level)* of 0.1 and a *log(Sb level)* of 0.05 has been used. Since 0.24 is closer to 0 we might conclude that it is unlikely that there will be a gold deposit within 0.5 km.

6.3.4 Significance of Model and Coefficients

When the coefficients of a logistic regression model are calculated, the method for identifying these coefficients attempts to maximize the *log likelihood* (*L*), which takes into account the difference between the actual value of *y* (y_i) and the predicted value π_i:

$$L = \sum_{i=1}^{n} (y_i \ln(\pi_i) + (1 - y_i) \ln(1 - \ln(\pi_i)))$$

In the Gold example, the log likelihood function is calculated to be −9.02. To better understand the significance of this model, this value is compared to the log likelihood for the constant only model (in this example −43.7). The model's log likelihood is clearly higher; however, we can perform a test using the likelihood ratio statistic (LR). This statistic takes into account both the *full* model (based on the two independent variables in this example) as well as the *reduced* model (in this case the constant only model). It is calculated using the following formula:

$$LR = -2 [L \,(\text{reduced}) - L \,(\text{full})]$$

In this example this translates to

$$LR = -2 [(-43.7) - (-9.02)] = 69$$

The LR statistic follows a chi-square distribution and we can determine a *p*-value by looking up the value in a chi-square distribution table where the degrees of freedom are the difference in the number of independent variables between the two models being assessed. In this case, $p < 0.0001$ and hence it would be considered significant. The LR statistic can be also used to understand the difference between two models containing different sets of independent variables.

We can also look at the significance of the individual coefficients in a manner similar to the linear regression coefficients. This assessment makes use of the Wald test (see Hosmer et al. (2013) for details). In this example, the coefficients are summarized in Table 6.7. It can be seen that

TABLE 6.7   Summary of the Logistic Regression Coefficients for the Gold Model

Independent Variable	Beta Coefficients	Standard Error	*z*-Values	*p*-Values
Constant	−1.84	0.95	−1.93	0.052
Log(As Level)	5.2	1.91	2.72	0.0063
Log(Sb Level)	3.5	1.82	1.92	0.054

TABLE 6.8 Accuracy, Sensitivity, and Specificity Based on Three Cut-Off Values

Cut-Off	Accuracy (%)	Sensitivity (%)	Specificity (%)
0.2	89.10	96.40	83.30
0.5	92.20	92.90	91.70
0.8	84.40	71.40	94.40

log(As level) has a low *p*-value below 0.05 and is clearly an important variable in the model; whereas *log(Sb level)* is around 0.05.

6.3.5 Classification

Up to this point we have been calculating a value for the response $E(Y|x)$; however, we may wish to have the predicted value in the same form as the *y*-variable, which is either 0 or 1. To map a value to either 0 or 1, we need to determine a cut-off value where values greater than or equal to the cut-off are assigned the value 1 and those less than the cut-off are assigned the value 0. Once we have the prediction in this form (0 or 1), we can assess the overall accuracy of the model in terms of the percentage of correct predictions as well as the sensitivity and specificity of the model.

One approach to determining the cut-off is to select a value by hand. To illustrate, we will use three possible cut-off values: 0.2, 0.5, and 0.8. By applying these values to the gold deposit example, the overall classification accuracy as well as sensitivity and specificity will change, as shown in Table 6.8. It is possible to automatically identify a cut-off value that is a good balance of sensitivity and specificity, and in this example it would be 0.532.

6.4 *k*-NEAREST NEIGHBORS

6.4.1 Overview

The *k-nearest neighbors* (*kNN*) method provides a simple approach to calculating predictions for unknown observations. This method calculates a prediction by looking at similar observations in the training set and uses some function of their response values, such as an average, to calculate the prediction. Like all prediction methods, it starts with a training set. It differs from other methods by determining the optimal number of similar observations to use in making the prediction rather than producing a mathematical model.

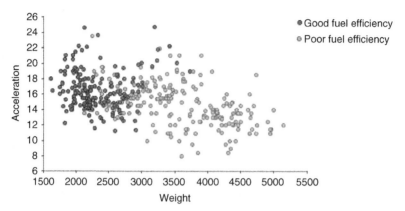

FIGURE 6.11 Scatterplot showing fuel efficiency classifications.

The scatterplot in Figure 6.11 is based on a data set of cars and will be used to illustrate how kNN operates. Two variables that will be used as independent variables are plotted on the *x*- and *y*-axis (*Weight* and *Acceleration*). The response variable is a dichotomous variable (*Fuel Efficiency*) which has two values: good and poor fuel efficiency. The darker shaded observations have good fuel efficiency and the lighter shaded observations have poor fuel efficiency.

During the learning phase, the "best" number of similar observations is chosen (*k*). The selection of *k* is described in Section 6.4.2. Once a value for *k* has been determined it is possible to make a prediction for a car with unknown fuel efficiency. To illustrate, cars A and B with unknown fuel efficiency are presented to the kNN model in Figure 6.12. The *Acceleration* and *Weight* of these observations are known and the two observations are plotted alongside the training set. Based on the optimal value for *k*, the *k*-nearest neighbors (most similar observations) to A and B are identified in Figure 6.12. For example, if *k* was calculated to be 10, then the 10 most similar observations from the training set would be selected. A prediction is made for A and B based on the response of the nearest neighbors (see Figure 6.13). In this case, observation A would be predicted to have good fuel efficiency since its neighbors primarily have good fuel efficiency. Observation B would be predicted to have poor fuel efficiency since its neighbors all have poor fuel efficiency.

kNN is relatively insensitive to errors or outliers in the data and can be used with large training sets; however, it can be computationally slow when it is applied to a new data set since a similarity score must be generated between the observations presented to the model and every member of the training set.

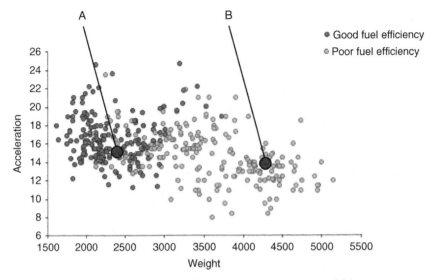

FIGURE 6.12   Two test set observations (A and B).

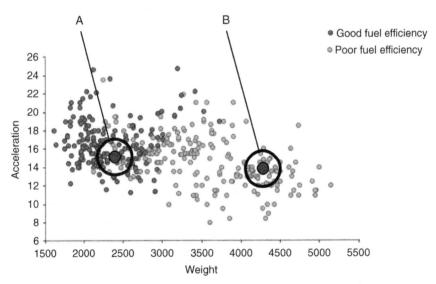

FIGURE 6.13   Looking at similar observation to support making predictions for A and B.

6.4.2 Training

A kNN model uses the k most similar neighbors to the observation to calculate a prediction. Where a response variable is continuous, the prediction is the mean of these nearest neighbors. Where a response variable is categorical, the prediction can be presented as a mean or a particular classification scheme that selects the most common classification term.

In the learning phase, three items should be considered and optimized: (1) the similarity method (2) the value of k, and (3) the combination of independent variables to use. As described in Chapter 5, there are many methods for determining whether two observations are similar including, for example, the *Euclidean* or the *Jaccard* distance. As with clustering, it is important to normalize the values of the variables so that no variables are considered to be more important based solely on the range of values over which they were measured. The number of similar observations that produces the best predictions or k must be determined. If this value is too high, the kNN model will overgeneralize; if the value is too small, it will lead to a large variation in the prediction.

The selection of k is performed by evaluating different values of k within a range and selecting the value that gives the "best" prediction. To ensure that models generated using different values of k are not over-fitting, a separate training and test set should be used, such as a 10% cross-validation.

To assess the different values for k, the *SSE* evaluation criteria will be used:

$$SSE = \sum_{i=1}^{n}(y_i - \hat{y}_i)^2$$

Smaller SSE values indicate that the predictions are closer to the actual values.

To illustrate, a data set of cars will be used and a model built to test the car's fuel efficiency (*MPG*). The following variables will be used as independent variables within the model: *Cylinders*, *Weight*, and *Acceleration*. The Euclidean distance calculation was selected to represent the distance between observations. To calculate an optimal value for k, different values of k were selected between 1 and 20. The SSE evaluation criterion was used to assess the quality of each model. In this example, the value of k with the lowest SSE value is 8 and this value is selected for use with the kNN model (see Table 6.9).

6.4.3 Predicting

Once a value for k has been set in the training phase, the model can now be used to make predictions. For example, an observation x has values for the

TABLE 6.9   Values for SSE for Different Values of *K*

k	SSE
1	12,003
2	8,358
3	7,525
4	7,246
5	6,870
6	6,906
7	6,628
8	**6,504**
9	6,533
10	6,658
11	6,621
12	6,612
13	6,648
14	6,773
15	6,811
16	6,943
17	6,965
18	7,015
19	6,963
20	6,996

independent variables but not for the response. Using the same technique for determining similarity as used in the model building phase, observation x is compared against all observations in the training set. A distance is computed between x and each training set observation. The closest k observations are selected and a prediction is made, for example, the average of the k-nearest neighbors is calculated to determine a prediction.

The observation (*Chevrolet Chevelle Malibu*) in Table 6.10 was presented to the kNN model built to predict a car's fuel efficiency (*MPG*). The *Chevrolet Chevelle Malibu* observation was compared to all observations in the training set and a Euclidean distance was computed. The eight observations with the smallest distance scores are selected, as shown

TABLE 6.10   kNN Test Observations

Car Name	Cylinders	Weight	Acceleration
Chevrolet Chevelle Malibu	6	3897	18.5

TABLE 6.11   Calculating the kNN Prediction Based on the Eight Nearest Neighbors

Car Name	Calculated Distance	Cylinders	Weight	Acceleration	MPG
Dodge Aspen	0.0794	6	3620	18.7	18.6
Amc Matador	0.0808	6	3632	18	16
Dodge Aspen Se	0.0844	6	3651	17.7	20
Plymouth Volare Custom	0.0894	6	3630	17.7	19
Chevrolet Chevelle Malibu Classic	0.0951	6	3781	17	16
Mercedes-Benz 280s	0.1093	6	3820	16.7	16.5
Ford Granada	0.1095	6	3525	19	18.5
Pontiac Phoenix Lj	0.1108	6	3535	19.2	19.2
				Average	18

in Table 6.11. The prediction is 18, which is the average of these eight observations.

The training set of observations can be used to explain how the prediction was reached and to assess the confidence in this prediction. For example, if the response values for all these observations are close, it increases the confidence in the prediction.

6.5 CLASSIFICATION AND REGRESSION TREES

6.5.1 Overview

In Chapter 5, decision trees were described as a way of grouping observations based on specific values or ranges of independent variables. For example, the tree in Figure 6.14 organizes a set of observations based on the car's number of cylinders (*Cylinders*). The tree was constructed using the variable *MPG* as the response variable. This variable was used to guide how the tree was constructed, resulting in groupings that characterize a car's fuel efficiency. The terminal nodes of the tree (A, B, and C) show a partitioning of cars into sets with good (node A), moderate (node B), and poor (node C) fuel efficiency.

Each terminal node is a mutually exclusive set of observations, that is, there is no overlap in observations between nodes A, B, or C. The criteria for inclusion in each of these nodes are defined by the set of branch points used to partition the data. For example, terminal node B is defined as observations where Cylinders ≥ 5 and Cylinders < 7.

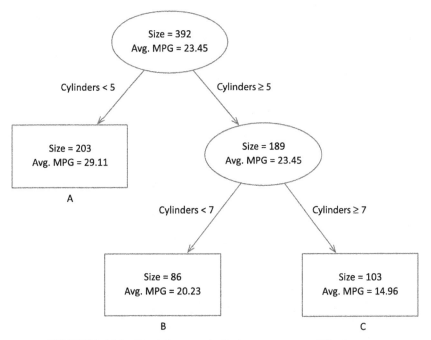

FIGURE 6.14   Decision tree built from an automobile data set.

Decision trees can be used as both classification and regression prediction models. Decision trees that are built to predict a continuous response variable are called *regression trees* and decision trees built to predict a categorical response are called *classification trees*. During the learning phase, a decision tree is constructed using the training set. Predictions in decision trees are made using the criteria associated with the trees terminal nodes. A new observation is assigned to a terminal node in the tree using these splitting criteria. The prediction for the new observation is either the node's classification (in the case of a classification tree) or the average value (in the case of a regression tree). As with other approaches to predtive modeling, the quality of the prediction can be assessed using a separate training set.

6.5.2 Predicting

In Figure 6.15, a set of cars are shown on a scatterplot. The cars are defined as having good or poor fuel efficiency. Those with good fuel efficiency are shaded darker than those with poor fuel efficiency. Values for the *Acceleration* and *Weight* variables are shown on the two axes.

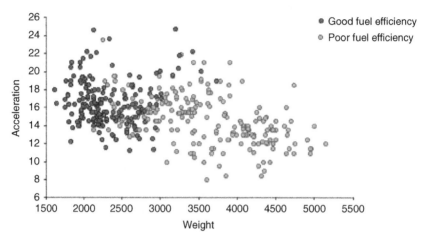

FIGURE 6.15   Scatterplot of cars show those with good and poor fuel efficiency.

A decision tree is generated using the car's fuel efficiency as the response variable. This results in a decision tree where the terminal nodes partition the set of observations according to ranges in the independent variables. One potential partition of the data is shown in Figure 6.16. The prediction is then made based on the observations used to train the model that are within the specific region, such as the most popular class or the average value (see Figure 6.17).

When an observation with unknown fuel efficiency is presented to the decision tree model it is placed within one of the regions. The placement is based on the observation's values for the independent variables. Two

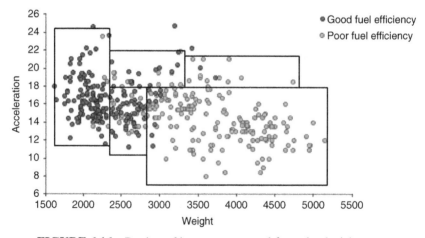

FIGURE 6.16   Region of interest generated from the decision tree.

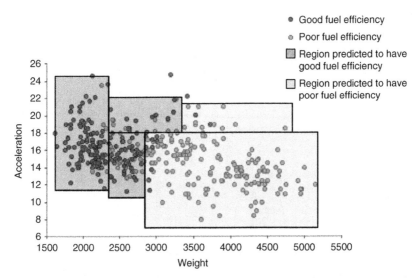

FIGURE 6.17   Scatterplot illustrating the classification response per region identified by the decision tree.

observations (A and B) with values for *Acceleration* and *Weight* but no value for whether the cars have a good or poor fuel efficiency are presented to the model. These observations are shown on the scatterplot in Figure 6.18 showing how the ranges of the variables used as independent variables partition the data. Observation A will be predicted to have good

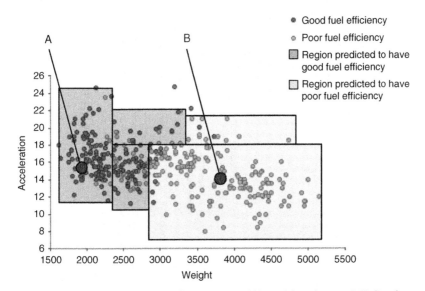

FIGURE 6.18   Classification of two automobiles with unknown MPG values.

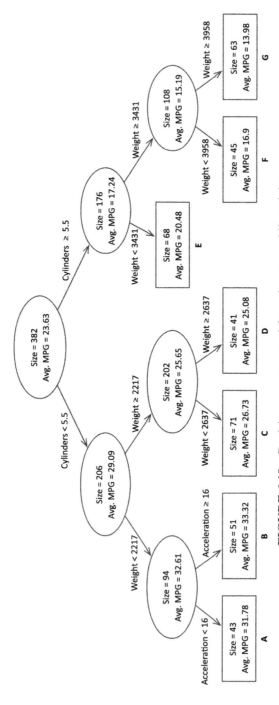

FIGURE 6.19 Decision tree generated from the automobile training set.

fuel efficiency whereas observation B will be predicted to have poor fuel efficiency.

Decision trees are useful for prediction since the results are easy to explain. Unfortunately these types of models can be quite sensitive to large variation in the training set that cannot be explained.

The same parameters used to build the tree (described in Section 5.5) can be set to build a decision tree model, that is, different input variable combinations and different stopping criteria for the tree.

6.5.3 Example

The decision tree in Figure 6.19 was built from a data set of 382 cars using the continuous variable *MPG* to split the observations. The average value shown in the diagram is the *MPG* value for the set of observations. The nodes are not split further if there are less than 40 observations in the terminal node.

In Table 6.12, three observations not used in building the tree are shown with both an actual and a predicted value. The second to last column (*Rule*) indicates the node in the tree that was used to calculate the prediction. For example, the "AMC Matador" with a weight of 3,730, six cylinders, and an acceleration of 19 will fit into a region defined by node F in the tree. Node F has an average *MPG* value of 16.9 and hence this is the predicted *MPG* value. The table also indicates the actual *MPG* values for the cars tested.

The examples used in this section were simple in order to describe how predictions can be made using decision trees. It is standard practice to use larger numbers of independent variables. Also, building a series of models by changing the terminating criteria can also be a useful way to optimize the decision tree models.

TABLE 6.12 Test Set of Three Automobiles Not Used in Building the Model

Car Name	Cylinders	Weight	Acceleration	MPG	Rule	Prediction
Oldsmobile Cutlass Salon Brougham	8	3420	22.2	23.9	E	20.48
Amc Matador	6	3730	19	15	F	16.9
Dodge D200	8	4382	13.5	11	G	13.98

The terminal nodes in the decision trees can be described as rules (as shown in Section 5.4.4) and these rules can be useful in explaining how a prediction was made. In addition, looking at the data on which each rule is based allows you to understand the degree of confidence for each prediction. For example, the number of observations and the distribution of the response variable can help to understand how much confidence you should have in the prediction.

6.6 OTHER APPROACHES

6.6.1 Neural Networks

Like all prediction models, the *neural network* approach uses a training set of examples to generate the model. This training set is used to generalize the relationships between the "input" independent variables and the "output" response variables. As part of this process, a series of interconnected nodes are organized between the input nodes (each input node corresponds to an independent variable) and the output response variables (represented as nodes). These nodes can be organized into multiple layers of interconnected nodes. As part of the process of learning, weights on the connections between the nodes in the neural network are refined to generate a prediction model. Once a neural network has been created, it can be used to make predictions. A prediction is made by presenting a test observation to the input nodes of the network and, based on local calculations, allowing the values to propagate through the network to eventually generate the scores for the output variables. These scores are the final prediction. Neural networks are a flexible way to generate models from the data and are capable of modeling complex linear and nonlinear relationships between the input variables and one or more response variables. They are, however, considered a "black box" since it is difficult to get a useful explanation of how the predictions were derived.

6.6.2 Support Vector Machines

Support vector machines are used primarily for classification problems. They attempt to identify a hyper-plane that separates the different classes being modeled. The observations on one side of the plane represent one of the classes being predicted, whereas the observations on the other side represent the other class. Despite their general usefulness and ability to handle complex classification problems, they can be difficult to interpret.

6.6.3 Discriminant Analysis

Discriminant analysis is an example of a classification approach. It classifies two or more values by constructing a linear combination of variables that characterize the different classes. One necessary assumption is that the independent variables are normally distributed.

6.6.4 Naïve Bayes

Naïve Bayes is a classification approach to building a predictive model. It makes use of the Bayesian theorem to compute probabilities of class membership. It provides a simple approach to modeling and can be easily used on large data sets. In addition, it can also be used to rank observations using a computed probability.

6.6.5 Random Forests

Random forests make use of multiple decision trees, with each tree using a different set of independent variables. The final prediction is calculated from the collection of decision tree results. As with Classification and Regression Trees or CART, this approach can be used for both classification and regression problems.

EXERCISES

1. A classification prediction model was built using a training set of examples. A separate test set of 20 examples is used to test the model. Table 6.13 shows the results of applying this test set.
   Calculate the model's
   (a) concordance
   (b) sensitivity
   (c) specificity

2. A regression prediction model was built using a training set of examples. A separate test set was applied to the model and the results are shown in Table 6.14.
   (a) Calculate the residual for each observation.
   (b) Determine the sum of the square of the residual.

3. Table 6.15 shows the relationship between the amount of fertilizer used and the height of a plant.

TABLE 6.13 Table of Actual Versus Predicted Values (Categorical Response)

Observation	Actual	Predicted
1	0	0
2	1	1
3	1	1
4	0	0
5	0	0
6	1	0
7	0	0
8	0	0
9	1	1
10	1	1
11	1	1
12	0	1
13	0	0
14	1	1
15	0	0
16	1	1
17	0	0
18	1	1
19	0	1
20	0	0

(a) Calculate a simple linear regression equation using *Fertilizer* as the independent variable and *Height* as the response.

(b) Predict the height when fertilizer is 12.3.

4. A kNN model is being used to predict house prices. A training set was used to generate a kNN model and k is determined to be 5. The unseen observation in Table 6.16 is presented to the model. The kNN model determines the five observations in Table 6.17 from the training set to be the most similar. What would be the predicted house price value?

5. A classification tree model is being used to predict which brand of printer a customer would purchase with a computer. The tree in Figure 6.20 was built from a training set of examples. For a customer whose Age = 32 and Income = \$35,000, which brand of printer would the tree predict they would buy?

TABLE 6.14   Table of Actual Versus Predicted (Continuous Response)

Observation	Actual	Predicted
1	13.7	12.4
2	17.5	16.1
3	8.4	6.7
4	16.2	15.7
5	5.6	8.4
6	20.4	15.6
7	12.7	13.5
8	5.9	6.4
9	18.5	15.4
10	17.2	14.5
11	5.9	5.1
12	9.4	10.2
13	14.8	12.5
14	5.8	5.4
15	12.5	13.6
16	10.4	11.8
17	8.9	7.2
18	12.5	11.2
19	18.5	17.4
20	11.7	12.5

TABLE 6.15   Table of Plant Experiment

Fertilizer	Height
10	0.7
5	0.4
12	0.8
18	1.4
14	1.1
7	0.6
15	1.3
13	1.1
6	0.6
8	0.7
9	0.7
11	0.9
16	1.3
20	1.5
17	1.3

TABLE 6.16   House with Unknown Price

Bedroom	Number of Bathrooms	Square Feet	Garage	House Price
2	2	1810	0	

TABLE 6.17   Table of Similar Observations

Bedroom	Number of Bathrooms	Square Feet	Garage	House Price
2	2	1504	0	355,000
2	2	1690	0	352,000
2	3	1945	0	349,000
3	2	2146	0	356,000
3	2	1942	0	351,000

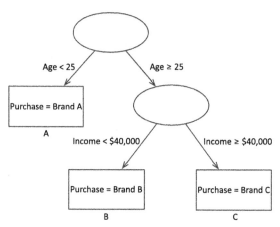

FIGURE 6.20   A classification tree model used to predict which brand of printer a customer would purchase.

FURTHER READING

When using statistical approaches (such as a t-tests) to identify independent variables from a large number of potential variables, it is important to use correction factors such as those discussed in Westfall et al. (1999) and Hsu (1996). Principal component analysis is an approach to dimension reduction and outlined in Jolliffe (2002) and Jackson (2003). See Kleinbaum et al. (2013) for a more detailed treatment of linear regression including a discussion of confidence intervals for coefficients and prediction; the use of the Durbin–Watson and autocorrelation methods for testing normality assumptions; and the use of automated approaches, including stepwise linear regression, to automatically identify combination of independent variables to use in the model. For a more detailed discussion on logistic regression, including the use of the Score Test, the use of the odds ratios to help interpret the models and the use of stepwise logistic regression see Hosmer et al. (2013). Agresti (2013) covers logistic regression as well as other methods to handle categorical data. For a more comprehensive treatment of advanced data mining approaches see Fausett (1993), Cristianini & Shawe-Taylor (2000), and Hastie et al. (2009).

APPENDIX A

ANSWERS TO EXERCISES

Chapter 2

1a. Nominal
1b. Ratio
1e. Ratio
1f. Ratio
1g. Ratio
1h. Ratio
1i. Nominal
2a. 45
2b. 45
2c. 48.7
2d. 53
2e. 324.9
2f. 18.02
 3. See Figure A.1

Making Sense of Data I: A Practical Guide to Exploratory Data Analysis and Data Mining,
Second Edition. Glenn J. Myatt and Wayne P. Johnson.

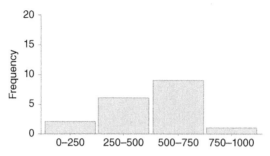

FIGURE A.1 Frequency distribution for Exercise 3 from Chapter 2.

Chapter 3

1. See Table A.1
2. See Table A.2
3. See Table A.3

Chapter 4

See Table A.4
2a. See Table A.5
2b. See Table A.6
2c. See Table A.7
See Figure A.2

TABLE A.1   Chapter 3, Question 1 Answer

Name	Weight (kg)	Weight (kg)—Normalize to 0–1
P. Lee	50	0.095
R. Jones	115	0.779
J. Smith	96	0.579
A. Patel	41	0
M. Owen	79	0.4
S. Green	109	0.716
N. Cook	73	0.337
W. Hands	104	0.663
P. Rice	64	0.242
F. Marsh	136	1

TABLE A.2   **Chapter 3, Question 2 Answer**

Name	Weight (kg)	Weight (kg)—Categorized (Low, Medium, High)
P. Lee	50	Low
R. Jones	115	High
J. Smith	96	Medium
A. Patel	41	Low
M. Owen	79	Medium
S. Green	109	High
N. Cook	73	Medium
W. Hands	104	High
P. Rice	64	Low
F. Marsh	136	High

TABLE A.3   **Chapter 3, Question 3 Answer**

Name	Weight (kg)	Height (m)	BMI
P. Lee	50	1.52	21.6
R. Jones	115	1.77	36.7
J. Smith	96	1.83	28.7
A. Patel	41	1.55	17.1
M. Owen	79	1.82	23.8
S. Green	109	1.89	30.5
N. Cook	73	1.76	23.6
W. Hands	104	1.71	35.6
P. Rice	64	1.74	21.1
F. Marsh	136	1.78	42.9

TABLE A.4   **Chapter 4, Question 1 Answer**

		Store		
		New York, NY	Washington, DC	Totals
Product Category	Laptop	1	2	3
	Printer	2	2	4
	Scanner	4	2	6
	Desktop	3	2	5
	Totals	10	8	18

TABLE A.5   Chapter 4, Question 2a Answer

Customer	Number of Observations	Sum of Sales Price ($)
B. March	3	1700
J. Bain	1	500
T. Goss	2	750
L. Nye	2	900
S. Cann	1	600
E. Sims	1	700
P. Judd	2	900
G. Hinton	4	2150
H. Fu	1	450
H. Taylor	1	400

TABLE A.6   Chapter 4, Question 2b Answer

Store	Number of Observations	Mean Sale Price ($)
New York, NY	10	485
Washington, DC	8	525

TABLE A.7   Chapter 4, Question 2c Answer

Product Category	Number of Observations	Sum of Profit ($)
Laptop	3	470
Printer	4	360
Scanner	6	640
Desktop	5	295

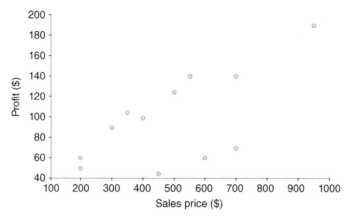

FIGURE A.2   Scatterplot, Chapter 4 question 3.

Chapter 5

1a. 4.8
1b. 2.8
1c. 0
2. 2.24
3. Support = 0.47, Confidence = 1, Lift = 1.89
4a. 0.83
4b. 0.998

Chapter 6

1a. 0.85
1b. 0.89
1c. 0.82
2a. 0.87

TABLE A.8   Chapter 6, Question 2b Answer

Observation	Actual	Predicted	Residual
1	13.7	12.4	1.3
2	17.5	16.1	1.4
3	8.4	6.7	1.7
4	16.2	15.7	0.5
5	5.6	8.4	−2.8
6	20.4	15.6	4.8
7	12.7	13.5	−0.8
8	5.9	6.4	−0.5
9	18.5	15.4	3.1
10	17.2	14.5	2.7
11	5.9	5.1	0.8
12	9.4	10.2	−0.8
13	14.8	12.5	2.3
14	5.8	5.4	0.4
15	12.5	13.6	−1.1
16	10.4	11.8	−1.4
17	8.9	7.2	1.7
18	12.5	11.2	1.3
19	18.5	17.4	1.1
20	11.7	12.5	−0.8

2b See Table A.8
3a. Height = −9.8 + 0.9 Fertilizer
3b. 1.3
4. $352,600
5. Brand B

APPENDIX B

HANDS-ON TUTORIALS

B.1 TUTORIAL OVERVIEW

Traceis 2014 is a software tool for exploratory data analysis and data mining, designed to be used alongside this book to provide practical experience of the methods described. It includes a number of tools for preparing and summarizing data, as well as methods for grouping, exploring patterns and trends, and building models. The following sections describe how to install and use the Traceis 2014 software and provide a series of hands-on exercises making use of sample data sets.

B.2 ACCESS AND INSTALLATION

The Traceis 2014 software can be accessed from the website http://www.makingsenseofdata.com/. The software is contained in a zipped file. Once downloaded, it can be unzipped into a folder on a computer. In addition to downloading the zipped file, a license key to use the software can be obtained by sending an email to software@makingsenseofdata.com. An email will be sent to you containing the key, which is simply a number.

Making Sense of Data I: A Practical Guide to Exploratory Data Analysis and Data Mining,
Second Edition. Glenn J. Myatt and Wayne P. Johnson.
© 2014 John Wiley & Sons, Inc. Published 2014 by John Wiley & Sons, Inc.

The associated website (http://www.makingsenseofdata.com/) contains the current minimum requirements for running the software, which can be used on a computer with the Java Virtual Machine (JVM) installed. The JVM usually comes installed on most computers; however, it can be downloaded from the Java website (http://www.java.com/), if necessary.

To run the software, either double click on the Traceis.jar file or open the Traceis.jar file from the Java software platform software. The first time the software runs, you will be asked to enter the license key number mentioned in the email sent. Also contained in the folder with the software is a subfolder called "Tutorial data sets," which contains sample data sets to use with the software, along with descriptions of the data sets.

B.3 SOFTWARE OVERVIEW

The Traceis 2014 software contains a series of tools, such as multiple linear regression or clustering. These tools are available throughout the Traceis 2014 user interface. The user interface is divided into five areas, as shown in Figure B.1: "Categories," "Tools," "Options," "Results," and "Selected observations."

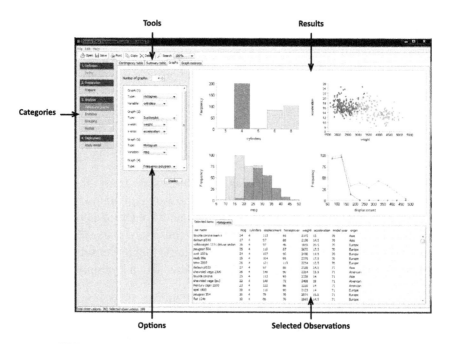

FIGURE B.1 Organization of the Traceis 2014 user interface.

TABLE B.1   Tools Available in the Traceis 2014 Software

Category	Tools
Preparation	Loading the data (open), searching the data set (search), characterizing variables (characterize), removing observations and variables (remove), cleaning the data (clean), transforming variables (transform), segmenting the data set (segment), and principal component analysis (PCA)
Tables and graphs	Contingency table, summary table, graphs, and graph matrices
Statistics	Descriptive statistics, confidence intervals, t-tests, chi-square test, ANOVA, and comparative statistics
Grouping	Clustering, association rules, decision trees
Models	Linear regression, discriminant analysis, logistic regression, Naïve Bayes, k-nearest neighbors (kNN), classification and regression trees (CART), and neural networks

The different types of tools are organized within the four-step process outlined in Chapter 1 of this book: (1) definition, (2) preparation, (3) analysis, and (4) deployment. The "Categories" options that can be selected are "Prepare," "Tables and graphs," "Statistics," "Grouping," "Models," and "Apply models." In Figure B.1 the "Tables and graphs" category has been selected and the tools available are presented accordingly, as shown in the tabs "Contingency table," "Summary table," "Graphs," and "Graph matrices." In this example, the "Graphs" tool was selected. The tools option area of the screen shows the different parameters and settings for performing or displaying an analysis of the data. For example, different types of graphs along with the graph's axes have been selected in Figure B.1. The "Results" area of the screen contains the results of a visualization or analysis. In Figure B.1 the selected graphs that are displayed are interactive. All cars with four cylinders were selected (by clicking on the four-cylinder bar in the top left chart) and the selected observations are highlighted on the other graphs in the display and shown in the selected observations area of the user interface. Table B.1 shows the tools that can be selected.

B.4   READING IN DATA

The first step is to load data into the system. The data set should be in a text file and should contain all observations in the data set with information on all variables. The format follows the conventions of comma- or tab-separated tabular data files. Each observation should be on a separate line and the values of the observed properties should be recorded consistently. The variable names should be on the first line in the file. A specific separator or delimiter should separate each individual value, such as a comma, tab, or semicolon. The following provides an example of the content of a text file containing a data table:

Ship Name;Cruise Line;Age;Tonnage;Passengers
Journey;Azamara;6;30.277;6.94
Quest;Azamara;6;30.277;6.94
Celebration;Carnival;26;47.262;14.86

The first row contains the column headings (variable names), each subsequent row contains the individual observations, and the values are consistently separated with semicolons.

Selecting the "Open" button, as shown in Figure B.1, initiates the process of loading a data set. Once the file has been located, you will be asked to review the data table to make sure it is formatted into the correct rows and columns, as shown in Figure B.2. If the data does not look correct,

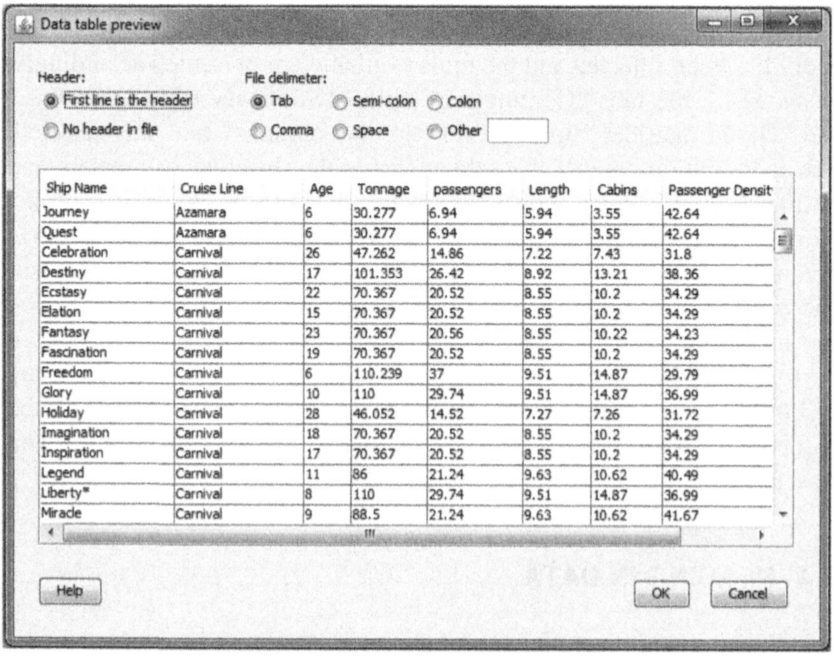

Data table preview

Header: ⦿ First line is the header   ◯ No header in file

File delimeter: ⦿ Tab ◯ Semi-colon ◯ Colon ◯ Comma ◯ Space ◯ Other [ ]

Ship Name	Cruise Line	Age	Tonnage	passengers	Length	Cabins	Passenger Densit
Journey	Azamara	6	30.277	6.94	5.94	3.55	42.64
Quest	Azamara	6	30.277	6.94	5.94	3.55	42.64
Celebration	Carnival	26	47.262	14.86	7.22	7.43	31.8
Destiny	Carnival	17	101.353	26.42	8.92	13.21	38.36
Ecstasy	Carnival	22	70.367	20.52	8.55	10.2	34.29
Elation	Carnival	15	70.367	20.52	8.55	10.2	34.29
Fantasy	Carnival	23	70.367	20.56	8.55	10.22	34.23
Fascination	Carnival	19	70.367	20.52	8.55	10.2	34.29
Freedom	Carnival	6	110.239	37	9.51	14.87	29.79
Glory	Carnival	10	110	29.74	9.51	14.87	36.99
Holiday	Carnival	28	46.052	14.52	7.27	7.26	31.72
Imagination	Carnival	18	70.367	20.52	8.55	10.2	34.29
Inspiration	Carnival	17	70.367	20.52	8.55	10.2	34.29
Legend	Carnival	11	86	21.24	9.63	10.62	40.49
Liberty"	Carnival	8	110	29.74	9.51	14.87	36.99
Miracle	Carnival	9	88.5	21.24	9.63	10.62	41.67

Help   OK   Cancel

FIGURE B.2 Preview of the data to be loaded.

you may override the default assumptions made by the software about the data format. If no header is contained in the text file, the software will automatically assign a header to each column ("Variable(1)," "Variable(2)," and so on). Once the data is correctly assigned as rows and columns, clicking on the "OK" button will load the data into the Traceis 2014 software.

B.5 PREPARATION TOOLS

B.5.1 Searching the Data

Once a data set has been loaded, one or more defined queries can be used to search the data set from the "Search" tab. These queries will search over specific variables. The queries may include different operators ($=$, $<$, $>$, and $\neq$), as well as specific values. You may also generate a new dummy variable from the search results, where a value of one is assigned if an observation satisfied the criteria of the search and zero is assigned otherwise. By checking the box "Generate variable from results" and entering a name, the new variable is generated after the search you initiate is completed.

B.5.2 Variable Characterization

Once loaded, the variables are analyzed and automatically assigned to various categories. This assignment can be reviewed by clicking on the "Characterization" tab.

B.5.3 Removing Observations and Variables

The "Remove" tab provides ways to remove observations or variables from the data table. Observations can be selected from most pages and may be removed by either clicking on the "Delete" button at the top of the user interface or from the "Remove" tab. Constants or specific variables can also be removed.

B.5.4 Cleaning the Data

For variables containing missing data or non-numeric data, a series of options are available from the "Clean" tab to "clean" the data. Once a variable is selected, the following summaries are provided: a count of the

TABLE B.2   Available Options for Transforming One or More Variables

Tool	Options
Normalization (new range)	Min-max, z-score, decimal scaling
Normalization (new distribution)	log, −log, Box-Cox
Normalization (text-to-numbers)	Values for selected variable
Value mapping (dummy variables)	Variables
Discretization (using ranges)	Select ranges for selected variable
Discretization (using values)	Select new values for existing values, for the selected variable
General transformations	Single variable: $x \times c$, $x + c$, $x \div c$, $c \div x$, $x - c$, $c - x$, x^2, x^3, and $\sqrt{x}$
	Two variables: mean, minimum, maximum, $x + y$, $x \div y$, $y \div x$, $x - y$, and $y - x$
	More than two variables: mean, minimum, maximum, sum, and product

numeric observations, a count of non-numeric observations, and a count of observations with missing data. Non-numeric observations in the data may either be removed or replaced by a numeric value. A similar set of options are available for handling missing data. Once the variable has been updated, the changes will be reflected in the results area of the updated table.

B.5.5 Transforming the Data

The "Transform" tab provides a number of options for transforming one or more variables into a new variable. These tools are summarized in Table B.2 and can be selected from the drop-down "Select type of transformation." It should be noted that all transformation options generate a new variable and do not replace the original variable(s).

The "Normalization (new range)" option provides three alternatives for transforming a single variable to a new range: min-max, z-score, and decimal scaling. Certain analysis options require that the frequency distribution reflect a normal or bell-shaped curve. The "Normalization (new distribution)" option provides a number of transformations that generates a new frequency distribution for a variable. The following transformations are available: log (log base 10 transformation), −log (a negative log base 10 transformation), and Box-Cox.

Certain ordinal variables contain text values, and before these variables can be used within numeric analyses a conversion from the text values to numeric values must be performed. The "Value mapping (text-to-number)" option provides tools to change each value of the selected variable into a specific number. To use nominal variables within numeric analyses, the variables are usually converted into a series of dummy variables. Each dummy variable corresponds to specific values for the nominal variable, where one indicates the presence of the value while a zero indicates its absence. The "Value mapping (dummy variables)" tool can be applied to a variable containing text values, and it will automatically generate a series of variables.

The "Discretization (using ranges)" option provides tools for converting a continuous numeric variable into a series of discrete values. Having selected a single variable, you can set the number of ranges and the lower and upper bounds for each range. Once the ranges are set, this tool substitutes the old, continuous numeric variable with the new value associated with the range in which that variable falls (greater than or equal to the lower bound and less than the upper bound). Additionally, categorical variables can be transformed to a series with fewer discrete values using "Discretization (using values)." Instead of grouping the values into ranges, this technique involves grouping the values of selected variables into a larger set, and assigning all of the observations within that larger set to the new value. The individual values can either be typed in or selected from a drop-down containing the alternatives already entered.

A series of "General transformations" can be selected in order to perform mathematical operations on one or more variables. Having selected "General transformation," one or more variables can be selected. To select a single variable, click once on the desired variable name; to select multiple variables use the <ctrl> + click to add individual items to the selection and <shift> + click to add all items between the current and the initial selection. When a single variable has been selected, the following common mathematical operations are available, where x refers to the selected variable and c is a specified constant: $x \times c, x + c, x \div c, c \div x, x - c, c - x, x^2, x^3$, and $\sqrt{x}$. When two variables x and y have been selected, in addition to the mean, minimum, and maximum functions, the following mathematical transformations are also available: $x + y, x \div y, y \div x, x - y$, and $y - x$. When more than two variables are selected, the following operations can be applied to the values: mean, minimum, maximum, sum, and product. By combining the results of these transformations, more complex formulas can be generated. Figure B.3 shows the generation of a specific new variable.

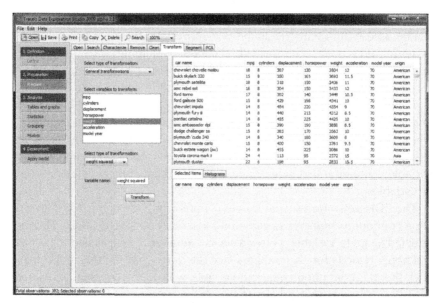

FIGURE B.3   A general transformation applied to a variable.

B.5.6 Segmentation

In some situations, to conduct an efficient analysis of the data it may be necessary to generate a smaller subset of observations. The desired number of observations to be included in the subset should be set from the "Segment" tab. The options "Random" and "Diverse" are available for selecting a subset of observations. The "Random" option will select the specified number of observations randomly and each observation in the original set has an equal chance of being included in the new set. The "Diverse" option will identify the desired number of observations that are representative of the original set. The selection of the diverse set of observations is done by clustering and then choosing from each cluster a representative for the cluster. The clustering is performed using k-means clustering, where k is the target number of observations in the subset, with the Euclidean distance metric measuring the distance between pairs of observations; the representative chosen from the cluster is the one closest to the center of the cluster. There are two options available from the "Select how to create subset" drop-down to describe what to do with the identified subset: (1) generate a new data set containing only the subset ("Remove observations"), or (2) generate a new dummy variable where one (1) represents the inclusion of the observation in the new subset, and zero (0) indicates its absence

("Generate dummy variable"). If option (2) is selected, the name of this new dummy variable should also be entered.

B.6 TABLES AND GRAPH TOOLS

B.6.1 Contingency Tables

A contingency table can be generated from the "Contingency table" option. The x- and y-axes on the table are both categorical variables, and they can be selected from the options. Having selected the x- and y-axes, clicking on the "Display" button will generate a contingency table in the results window. The table shows counts corresponding to pairs of values from the selected categorical variables. In addition, totals are presented for each row and column in the table. Each of the cells in the table can be selected, and the observations included in these counts will be displayed in the selected observations panel and highlighted in other views involving those observations. An example is shown in Figure B.4.

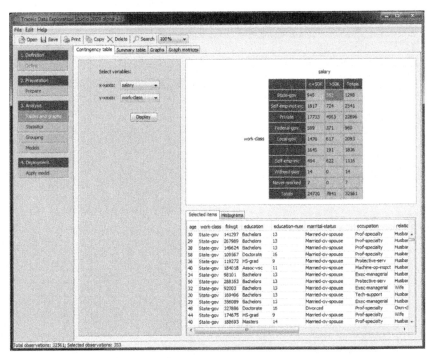

FIGURE B.4 Contingency table generated from a data set.

B.6.2 Summary Tables

A summary table groups observations using a single categorical variable and provides summarized information about other variables for each of these groups. A summary table can be generated from the "Summary table" tab. First, a categorical variable for grouping the observations is selected. An optional count of the number of observations in each group can also be selected. A number of additional columns can be added to the table, and this number is set with the "Number of columns" counter. There are seven options for summarizing each selected variable: (1) mean, (2) mode, (3) minimum, (4) maximum, (5) sum, (6) variance, and (7) standard deviation. Clicking on the "Display" button will generate a summary table corresponding to the options selected. Individual rows can be selected, and the resulting observations will be shown in the selected observations panel as well as in other views, as illustrated in Figure B.5.

B.6.3 Graphs

One or more graphs can be shown on a single screen to summarize the data set and these specific graphs are selected from the "Graphs" tab.

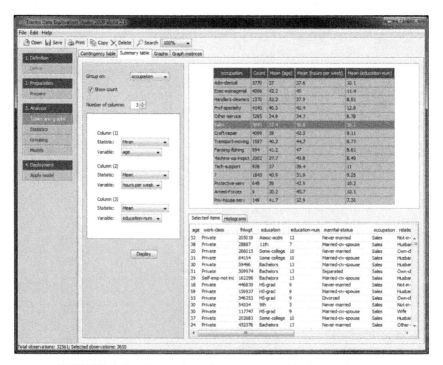

FIGURE B.5   Generation of a summary table from a data set.

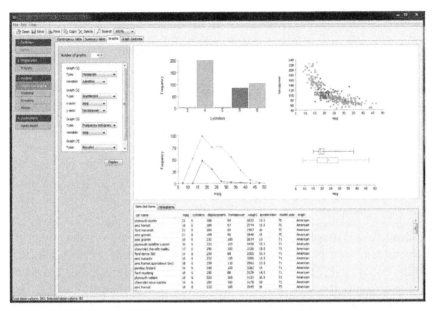

FIGURE B.6   Different data graphs generated from the data set.

After identifying the desired number of graphs to display, a series of options for each graph is provided. There are four types of graphs: (1) histogram, (2) scatterplot, (3) box plot, and (4) frequency polygram. In addition, the variable or pair of variables to display should be selected. Once the collection of graphs to display has been determined, clicking on the "Display" button will show these graphs in the results area. In each of the graphs, selected observations will be highlighted on all graphs with darker shading. The histograms, frequency polygrams, and scatterplots are all interactive. For instance, observations can be selected by clicking on a histogram bar or a point in the scatterplot or frequency polygram. In addition, a lasso can be drawn around multiple bars or points. Any selection will be updated on the other graphs in the results area, as well as being made available in other analysis options throughout the program, as illustrated in Figure B.6.

B.6.4 Graph Matrices

The "Graph matrices" tab generates a matrix of graphs in one view. The tools provide options to display a histogram, scatterplot, or a box plot matrix for the selected variables. Multiple variables can be selected using <ctrl> + click for non-contiguous variables, <shift> + click for continuous variables, and <ctrl> + A for all variables. Clicking on the "Display" button will show the matrix in the results area. The histogram and box plot matrices present the graphs only for the selected variables. In contrast, the scatterplot matrix shows scatterplots for all combinations of the variables selected. The names of the scatterplot axes are shown in the boxes where no graphs are drawn.

B.7 STATISTICS TOOLS

B.7.1 Descriptive Statistics

The tools available in the "Descriptive" tab will generate a variety of descriptive statistics for a single variable. For the selected variable, descriptive statistics can be generated for (1) all observations, (2) the selected observations, and (3) observations not selected. Clicking on the "Display" button presents the selected descriptive statistics in the results area. For each of the sets of observations selected, a number of descriptive statistics are calculated. They are organized into the following categories: number of observations, central tendency (mode, medium, and mean), variation (minimum value, maximum value, the three quartiles—Q1, Q2, and Q3, variance, and standard deviation), and shape (estimates of skewness and kurtosis).

B.7.2 Confidence Intervals

The "Confidence intervals" analysis calculates an interval estimate for a selected variable that is based on a specific confidence level. In addition, confidence intervals for groups of observations—defined using a single categorical variable—can also be displayed. The confidence intervals for the variables, as well as for any selected groups, can be seen in the results area.

B.7.3 *t*-test

The "*t*-test" tool will perform a hypothesis test on a single variable. When a categorical variable is selected, the hypothesis test takes into consideration the proportion of the selected variable with a specified value. This value must be set with the "Where x is:" drop-down option, where x is the selected categorical variable. When a continuous variable is selected, the hypothesis test uses the mean. The test can take into consideration one or two groups of observations. When the "single group" option is selected, the members of this group should be defined. The four alternatives for membership are: (1) all observations in the data table, (2) selected observations, (3) observations not selected, or (4) observations corresponding to a specific value of a categorical variable. The confidence level, or alpha, should be selected and set to one of the following: 0.1, 0.05, or 0.01. The hypothesis test should then be described for the selected observations. The value for the null hypothesis should be set along with information about whether the value for the alternative hypothesis should be greater than, less than, or not equal to value for the null hypothesis.

There are two options when looking at two groups: "Two groups (equal variance)" and "Two groups (unequal variance)." After the two groups are selected, the members of each group should then be defined. The three alternatives are: (1) selected observations, (2) observations not selected, and (3) observations corresponding to a specific value of a categorical variable. As before, the confidence level, or alpha, should be selected from one of the following: 0.1, 0.05, or 0.01. The specific hypothesis test should be defined for the selected observations. The choice for null hypothesis will be either that the two means are equal or the proportions of the two selected variables are the same. Again, the value of the alternative hypothesis should be set to either "less than," "greater than," or "not equal to" the value of the null hypothesis. The results of the hypothesis test are presented in the results area. These results include details of the variable and the observations assessed, including the mean value or values, the actual hypothesis test with confidence level, as well as the hypothesis or z-score, the critical z-score, the p-value, and whether to accept or reject the null hypothesis.

B.7.4 Chi-Square Test

The "Chi-Square" option allows for comparison between two categorical variables that are selected from the two drop-down menus. The results of the analysis are shown in two contingency tables in the results area. One of the contingency tables includes the actual data, and the second contains expected results. A Chi-Square assessment is also calculated along with its attained significance level.

B.7.5 ANOVA

The "ANOVA" tool assesses the relationship for a particular variable between different groups of observations within that variable. The variable to assess and the categorical variable used to group the observations

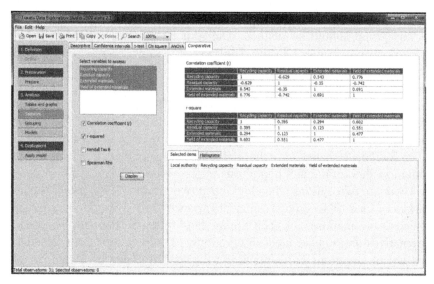

FIGURE B.7 Comparing variables within a data set.

should be selected and a confidence level, or alpha, should be assigned. The resulting analysis is presented in the results area, which shows the groups identified using the selected categorical variable, the number of observations in each group, the mean value for each group, and the variance for each group. The mean square within and between is calculated, along with the F-statistic and p-value.

B.7.6 Comparative Statistics

There are a number of ways to calculate metrics showing the strength of the relationships between combinations of variables. There are four options for displaying the association between variables: the correlation coefficient (r), the squared correlational coefficient (r^2), Kendall Tau, and Spearman Rho. For the selected variables the tool will present in one or more tables the select coefficients for all pairs of variables, as illustrated in Figure B.7.

B.8 GROUPING TOOLS

B.8.1 Clustering

A number of methods for clustering observations are available including agglomerative hierarchical clustering and partitioned (k-means) clustering. These clustering methods require the selection of one or more variables, as well as the selection of a distance measure used to assess the degree of similarity between observations. For numeric variables (not binary), a number of distance calculations are available including Euclidean; for binary variables, a different set of distance methods are provided including Jaccard. It should be noted that it is not necessary to normalize the data to a standard range, as the software will perform this step automatically.

If the "Agglomerative hierarchical clustering" option is selected, a linkage method must be chosen from the list: average linkage, complete linkage, or single linkage. The assignment of the observations to specific clusters can be stored as a separate column in the table by selecting the "Generate clusters as column(s)" option where a single variable will be generated. Having specified the type of clustering required, clicking on the "Cluster" button will generate the clusters and the results will be summarized in the results area. When agglomerative hierarchical clustering is selected, the results are displayed as a dendrogram showing the hierarchical organization of the data. A vertical line dissects the dendrogram, thus creating clusters of observations to the right of the vertical line. A rectangle is placed around each cluster and, space permitting, a number indicating the size of the cluster is annotated on the right. When the data set has a label variable, clusters with only a single observation are replaced by the label's value. The cut-off is interactive; it can be moved by clicking on the square toward the bottom of the cut-off line and moving it to the left or right. Moving the line changes the distance at which the clusters are generated, and hence the number of clusters will change as the cut-off line moves. If the "Generate clusters as column(s)" option is selected, the column in the data table describing the cluster membership will also change. Agglomerative hierarchical clustering is illustrated in Figure B.8.

If the "Partitioned (k-means)" clustering option is selected, the number of clusters needs to be specified and the results are presented in a table where each row represents a single cluster. The centroid values for each cluster are presented next to the number of observations in the cluster. This "number of observations" cell in the table can be selected, and those selected observations are displayed in the selected observations panel, as well as throughout the program.

B.8.2 Association Rules

The "Association rules" option will group observations into overlapping groups that are characterized by "If ... then ... " rules. A set of categorical variables can be selected, and specific values corresponding to

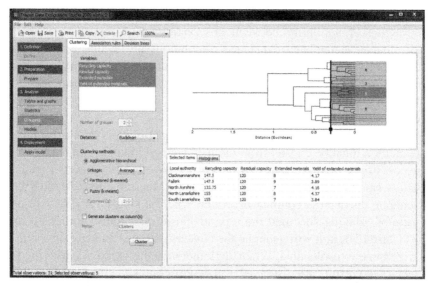

FIGURE B.8   Illustration of agglomerative hierarchical clustering.

these variables will be used in the generated rules. The software includes a "Restrict rules on the THEN-part" option, which will only result in rules where the THEN-part incorporates the selected variable. Also, the "Restrict rules by specific value" function allows for the selection of an appropriate value from the drop-down list. This option is particularly useful when the rules are being generated from a series of dummy variables, and only rules with values of "one" (1) contain useful information. Generating rules with a minimum value for support, consequence, and lift can also be set. The resulting rules are shown as a table in the results area, where the "IF-part" of the rule is shown in the first column, and the "THEN-part" of the rule is shown in the second column. Other columns display a count of the number of observations from which the rule is derived. The table also displays values for support, consequence, and lift. The table can be sorted by any of these columns. Selecting a single row will display the observations in the selected observations panel (as illustrated in Figure B.9), and those observations will be highlighted throughout the program.

B.8.3 Decision Trees

A decision tree can be built from a data table using the "Decision tree" option. Any number of variables can be used as independent variables and a single variable should be assigned as the response. In addition, a minimum

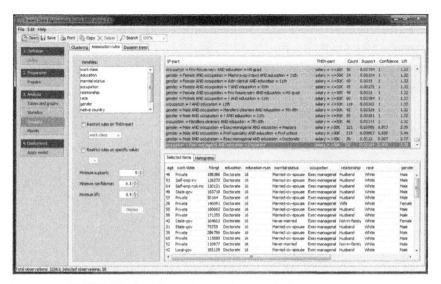

FIGURE B.9   Illustration of the association rules results.

tree node size should be set which prevents the software from generating a tree with fewer nodes than this specified value. Once a decision tree has been built, the results will be shown in the results area. The nodes of the decision tree represent sets of observations, which are summarized with a count, along with the average value for the response variable (if the response variable is continuous). For categorical response variables, the number of observations in the set is shown, along with the most common value qualified by the number of occurrences of that value compared to the total number of nodes. In both trees, the criteria used to split the trees are indicated just above the node. Oval nodes represent non-terminal nodes, whereas rectangular nodes represent terminal nodes. The decision trees are interactive; each node can be selected, and the selected observations can be seen below the tree as well as in other views, as illustrated in Figure B.10.

B.9 MODELS TOOLS

B.9.1 Linear Regression

Tools for building multiple linear regression models are found under the "Linear regression" tab. Any number of independent variables can be selected using <ctrl> + click to select non-contiguous individual variables and <shift> + click to select a continuous range of variables. A single

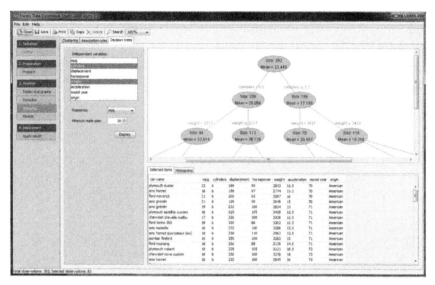

FIGURE B.10   Illustration of view with decision tree results.

continuous response variable should also be selected. The cross-validation percentage should be set to indicate the proportion of observations to set aside for testing. To further analyze the results, a series of new variables can be generated: *Prediction*, *CV Prediction*, and *Order*. A final multiple linear regression model will be built from the entire data set, and the *Prediction* variable will have a prediction of all observations from this model. *CV Prediction* represents the cross-validated prediction, where the predicted values are calculated using a model built from other observations. *Order* contains a number for each observation reflecting the order in which the observation appears in the data set.

Once a model is built, the results are displayed in the results area. The independent variables and the response variable are initially summarized, including the model coefficients and the significance of the coefficients. The software provides a regression analysis summarizing the model accuracy, including R-squared, adjusted R-squared, and standard error. An ANOVA analysis is generated showing the degrees of freedom (df) of the regression and the residual, along with the mean square (MS), the sum of squares (SS), the F-statistic, and the p-value (as shown in Figure B.11). To evaluate the model in more detail, a residual variable can be generated using the "Transform" tab option under "Preparation." To create a residual variable, first select a "General transformation" and select the actual response and the prediction, and then select "Actual–prediction." This data can be plotted in the "Graphs" tab to analyze the model further. Once a model is

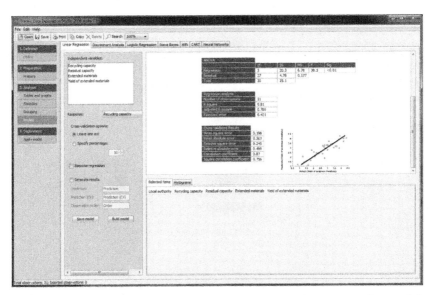

FIGURE B.11   Illustration of the results of running a linear regression model.

built and evaluated, it can be saved permanently. The model is saved by clicking on the "Save model" button. A file name should be provided and the model will be saved to a file for future use with other data sets.

B.9.2 Logistic Regression

The "Logistic regression" option enables the generation of a logistic regression model which can only be built for binary response variables. Numeric variables can be used as independent variables. A logistic regression model generates an expected value of the y-variable or probability, and the classification prediction is generated from this probability using a specified threshold value or by automatically generating a cut-off which is a good balance of sensitivity and specificity. Observations above this threshold value will be assigned a prediction of one, and those below will be assigned a prediction of zero. In addition to the cross-validated percentage to set aside, a number of predicted values can be optionally generated by selecting the "Generate results" option: "Prediction," "Prediction prob.," "CV Prediction," and "CV Prediction prob." The "Prediction" is the 0 or 1 classification using the final model with the "Prediction prob." being the probability calculated. "CV Prediction" and "CV Prediction prob." report the same information as part of the cross-validation. Once a model has been built, the results are displayed in the results area. The independent variables and the response variable are initially summarized, along with

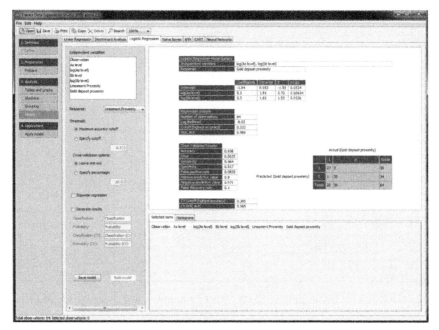

FIGURE B.12 Illustration of the results from a logistic regression.

the model coefficients and their significance. A summary of the cross-validated results is also presented. Figure B.12 illustrates the results from a logistic regression model. Generated models can be saved for use with other data sets using the "Save model."

B.9.3 *k*-Nearest Neighbor

The "*k*NN" analysis tab lets you build *k*-nearest neighbor models. Models can be built for any type of response variable, and any type of numeric variable can be selected as an independent variable. It is not necessary to normalize the data to a standard range as the software will do this automatically. The distance metric should be selected from the drop-down menu. A value for *k* can be set manually. Alternatively, a range can be specified which instructs the Traceis software to build all models between the lower and upper bound and from these select the model with the smallest error. In addition to the cross-validated percentage to set aside, a number of predicted values can be optionally generated by selecting the "Generate results" option: "*k*NN-Pred" and "*k*NN-Pred(CV)," which are the predicted values for the final model, along with the predicted values from the cross-validated models. Once a model has been built, the results are displayed in the results area. The independent variables and the response variable are shown, as well as the value for the k-nearest neighbor parameter that was either set manually or automatically derived. A summary of the standard cross-validated results is presented. Models that are generated can be saved for future use with other data sets, using the "Save model" button.

B.9.4 CART

The "CART" analysis tab is used to build models based on either a regression tree or a classification tree. Models can be built for any type of response variable or independent variables. Decision trees are generated for the models. If the "Minimum node size" is set, the process that generates decision tree will prune from the tree nodes that are smaller than the specified size. Other predicted values can be optionally generated by selecting the "Generate results" option: "CART-Pred" and "CART-Pred(CV)." Once a model has been built, the results are displayed in the results area. The independent variables along with the response variable are shown, as well as the values used for minimum node size, where size is a count of the number of observations at each node. A summary of the standard cross-validated results is presented. Models that are generated can be saved for later use with other data sets, using the "Save model" button.

B.10 APPLY MODEL

Models that are built and saved can be used with a new data set by selecting the "Apply model" option under the "4. Deployment" step. When the model and the new data set are opened, a summary of the model and the data is shown. Selecting the "Apply" button will generate a prediction for the observations in the new data set. The column headings must match those used to build the model. In addition, the ranges of the variables must be within the variable ranges used to build the model, otherwise a prediction cannot be generated.

B.11 EXERCISES

B.11.1 Overview

The following four exercises are available to use with the Traceis software or other available software. The data set for use with this hands-on tutorial are available from the tutorials folder. In addition, a series of other data sets is available to use with the Traceis 2014 software.

B.11.2 Exercise 1: Analysis of Recycling Data

The data set contains 31 observations, where each observation represents a Scottish local authority from Baird et al. (2013). The file contains five variables (1) *Local authority*, (2) *Recycling capacity* (in liters/week), (3) *Residual Capacity* (in liters/week), (4) *Extended materials* (number of extended materials collected), and (5) *Yield of extended materials* (in kg/hh/wk). The data set is available in the tutorials folder under the name *scottish_recycle.txt*.

The objective of this exercise is to generate a regression model that predicts the *Yield of extended materials*.

Step 1: Load the data set into Traceis 2014 software by clicking on the "Open" button and find the file *scottish_recycle.txt* in the tutorials folder. Select "OK" from the preview window and you should be able to see the data as in Figure B.13.

Step 2: Characterize the variables based on the scales over which they were measured.

Step 3: Plot the frequency histogram (from the "Graphs" panel) and generate summary statistics (from the "Descriptive" statistics panel) for each numeric variable, generating the information shown in Figure B.14.

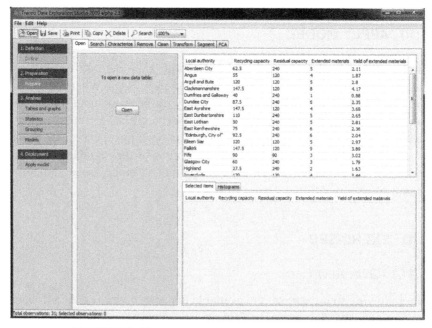

FIGURE B.13 Data loaded into the Traceis software.

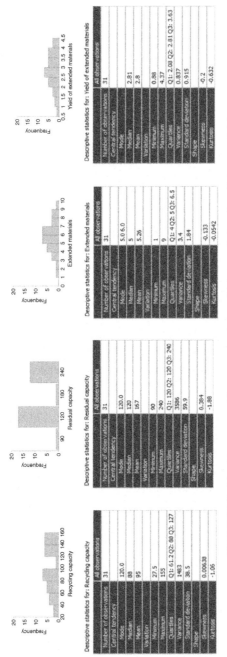

Descriptive statistics for: Recycling capacity

	All observations
Number of observations	31
Central tendency	
Mode	120.0
Median	88
Mean	95
Variation	
Minimum	27.5
Maximum	155
Quartiles	Q1: 61.2 Q2: 88 Q3: 127
Variance	1483
Standard deviation	38.5
Shape	
Skewness	0.006.38
Kurtosis	-1.06

Descriptive statistics for: Residual capacity

	All observations
Number of observations	31
Central tendency	
Mode	120.0
Median	120
Mean	167
Variation	
Minimum	90
Maximum	240
Quartiles	Q1: 120 Q2: 120 Q3: 240
Variance	3586
Standard deviation	59.9
Shape	
Skewness	0.384
Kurtosis	-1.88

Descriptive statistics for: Extended materials

	All observations
Number of observations	31
Central tendency	
Mode	5.0 6.0
Median	5
Mean	5.26
Variation	
Minimum	1
Maximum	9
Quartiles	Q1: 4 Q2: 5 Q3: 6.5
Variance	3.4
Standard deviation	1.84
Shape	
Skewness	-0.133
Kurtosis	-0.0542

Descriptive statistics for: Yield of extended materials

	All observations
Number of observations	31
Central tendency	
Mode	2.81
Median	2.8
Variation	
Minimum	0.88
Maximum	4.37
Quartiles	Q1: 2.08 Q2: 2.81 Q3: 3.63
Variance	0.837
Standard deviation	0.915
Shape	
Skewness	-0.2
Kurtosis	-0.632

FIGURE B.14 Frequency histogram and descriptive statistics for all four variables.

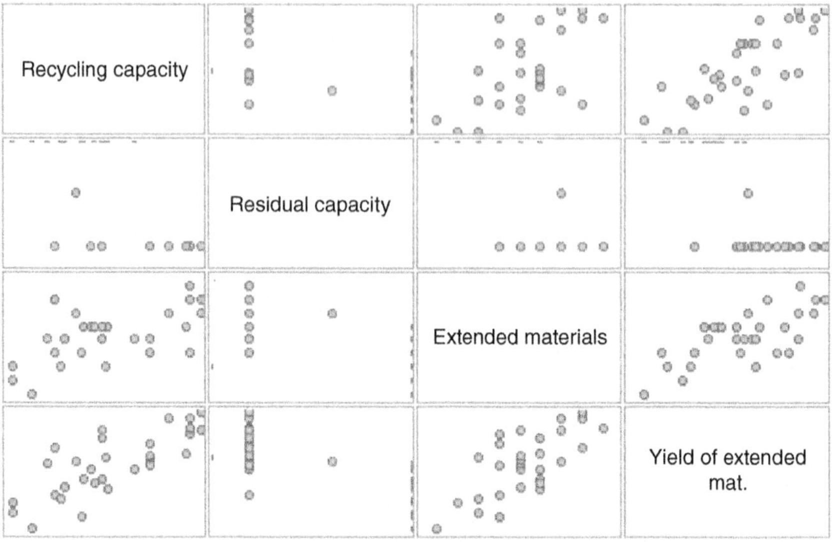

Correlation coefficient (r)	Recycling capacity	Residual capacity	Extended materials	Yield of extended materials
Recycling capacity	1	-0.629	0.543	0.776
Residual capacity	-0.629	1	-0.35	-0.742
Extended materials	0.543	-0.35	1	0.691
Yield of extended materials	0.776	-0.742	0.691	1

FIGURE B.15   Scatterplot matrix and correlation coefficients.

Step 4: Look at the relationship between the variables using the scatterplot matrix ("Graph matrices") and the correlation statistics (r) ("Comparative statistics"), as shown in Figure B.15.

Step 5: Build a series of linear regression models (from models "Linear Regression") using all combinations of independent variables (*Recycling capacity, Residual capacity*, and *Extended materials*) to predict *Yield of extended materials* and select the best performing, simplest model (as seen in Figure B.16). Check the "Generate results" option and click on "Save" to save this model.

Step 6: Calculate a residual value (from "Transform," which is an option under the "Prepare" category). Choose "General transformations," select the *Prediction* variable as well as the actual response variable (*Yield of extended materials*) using <ctrl> + click for non-contiguous variables selection and select the specific transformation of "Yield of extended material—Prediction." Name the new variable "Residual" as shown in Figure B.17. Look at the different graphs (from the "Graphs" tab), as shown in Figure B.18, to test the model assumptions.

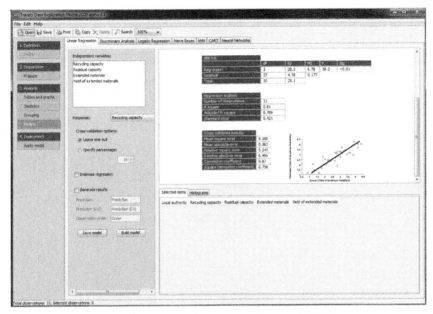

FIGURE B.16 Linear regression model built.

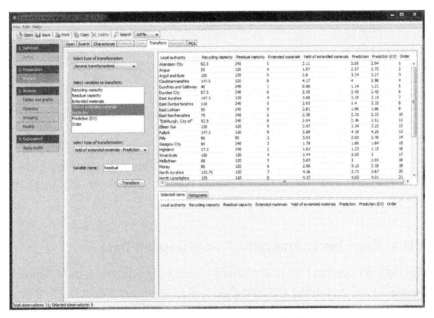

FIGURE B.17 Calculating a residual value.

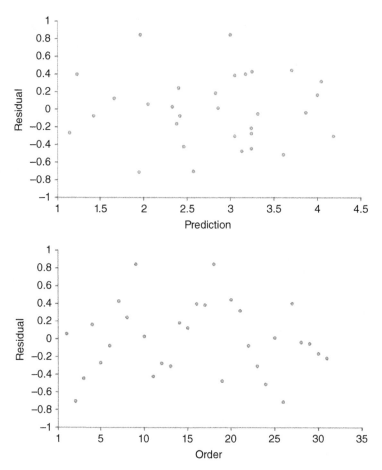

FIGURE B.18 Testing linear regression assumptions.

Step 7: Apply the saved model as shown in Figure B.19, using either a hypothetical test set with the same column names or the original data set.

B.11.3 Exercise 2: Analysis of Gold Deposit Data

This data set contains 64 observations concerning whether a gold deposit was identified within 0.5 km (*Gold deposit proximity*) (Sahoo & Pandalai, 1999). For this variable, the values are 1 if a gold deposit was identified and 0 if not. Other variables were measured that will be used to predict

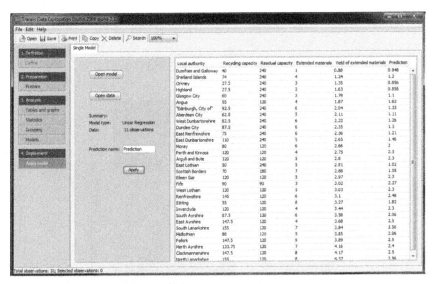

FIGURE B.19   Applying a saved model.

whether a gold deposit is within 0.5 km: *As level*, *Sb level*, and *Lineament proximity*.

The objective of this exercise is to develop a classification model to predict *Gold deposit proximity* from any combination of the collected data (*As level*, *Sb level*, *Lineament proximity*).

Step 1: Load the data into the Traceis software by selecting the *gold_target1.txt* from the tutorials folder.

Step 2: Explore and determine the scales over which the variables are measured.

Step 3: Look at the frequency distribution and the descriptive statistics for the variables in the data, as shown in Figure B.20, using the "Graphs" and "Descriptive" statistics tools.

Step 4: Since the *As level* and *Sb level* variables do not follow a normal distribution, perform a log transformation on the values of these two variables (from the "Transform" tool using "Normalization (new distribution)" option) and then re-examine the new frequency distributions (as illustrated in Figure B.21).

Step 5: Look at the relationships between *Gold deposit proximity* and *Lineaments proximity* using the "Contingency table" tool. Next, explore the relationship between *Gold deposit proximity* and *As level* and *Sb level* by first creating a discretized version of the *log (As level)* and *log*

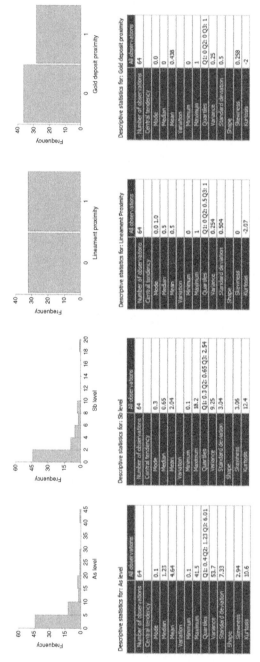

FIGURE B.20 Frequency distribution and descriptive statistics for four variables.

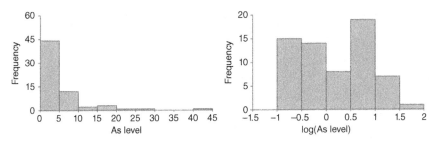

FIGURE B.21   Frequency distribution for As level and Sb level as well as log transformed As level and Sb level.

(Sb level) variables using the "Transform" tool with the "Discretization (using ranges)" options, and then using the "Summary table" tool to create a summary of how the mean *Gold deposit proximity* values change as the As and Sb levels increase. Figure B.22 illustrates the three graphs.

Step 6: Build a series of logistic regression models (from the "Logistic regression" tool) to predict *Gold deposit proximity* from all combinations of: *Lineament proximity*, *log (As level)*, and *log (Sb level)*, as illustrated in Figure B.23. Select the simplest, best performing model and save the model.

B.11.4 Exercise 3: Analysis of Morphologic Difference across the *Iris* Plant Species

This data set contains 150 observations concerning morphologic differences across three species of the *Iris* flower (*class*): "Iris setosa," "Iris virginica," and "Iris versicolor" from Fisher (1936). In biology, *morphology* refers to the study of living organisms through their form and structure.

The objective of this analysis is to understand how the three species are characterized by the morphologic variations: *sepal width (cm)*, *sepal length (cm)*, *petal width (cm)*, and *petal length (cm)*. These will be your morphological variables.

Step 1: Open the *IRIS.txt* file located in the tutorials directory.

discretized log (As level)	Count	Mean (Gold deposit proximity)
-1 -> -0.5	15	0
-0.5 -> 0	14	0.0714
0 -> 0.5	8	0.125
0.5 -> 1	19	0.947
1 -> 1.5	7	1
1.5 -> 2	1	1

discretized log (Sb level)	Count	Mean (Gold deposit proximity)
-1 -> -0.5	18	0
-0.5 -> 0	17	0.118
0 -> 0.5	15	0.867
0.5 -> 1	13	0.923
1 -> 1.5	1	1

Lineament proximity

	0	1	Totals
0	30	6	36
1	2	26	28
Totals	32	32	64

FIGURE B.22   Exploring the relationships to gold deposit proximity.

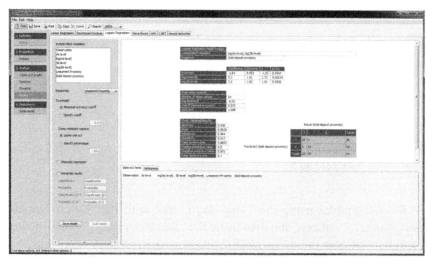

FIGURE B.23 Logistic regression model built to predict gold deposit proximity.

Step 2: Characterize the scales over which the variables are measured and calculate descriptive statistics and frequency distributions for all measured variables (as shown in Figure B.24).

Step 3: Generate three scatterplot matrices using the measured morphologic variables where each matrix is highlighted with a different species of the Iris flower, as shown in Figure B.25. The histogram of the Iris flower species can be displayed from the "Graphs" tool and highlighted by clicking on the histogram bar. The "Graph matrices" tool can be used to generate the scatterplot matrix where any selected set of observations will also be highlighted.

Step 4: Cluster the data set with agglomerative hierarchical clustering (using Euclidean distance to measure similarity between observations and average linkage to determine how clusters are formed) and set the cut-off value such that there are three clusters generated. Figure B.26 shows how the scatterplot matrices and the species histogram are highlighted when you select each of the three clusters (as shown in the first row of graphs in Figure B.26).

Step 5: Build a classification tree (with the "Decision tree" tool) using the four measured properties and the response (*class*) to guide how the tree was generated, using a minimum of 30 observations per node (as seen in Figure B.27).

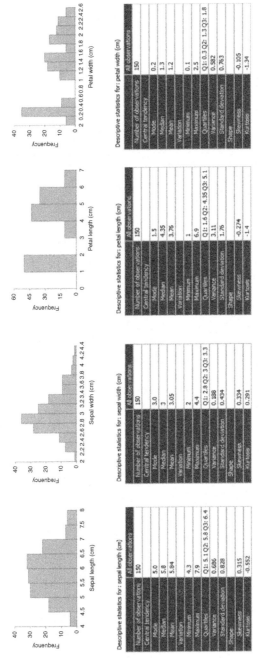

FIGURE B.24 Frequency histogram and descriptive statistics for all four measured variables from the *Iris* data set.

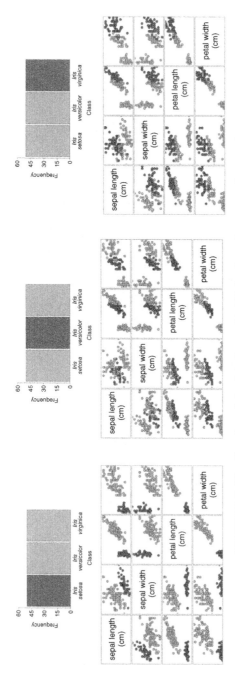

FIGURE B.25 Species of the *Iris* flower highlighted on the scatterplot matrix.

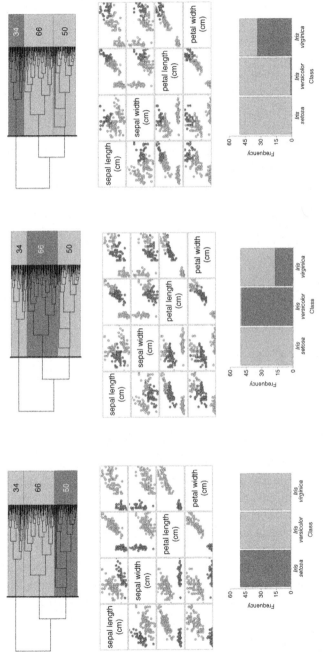

FIGURE B.26 Visualization of the clustering results.

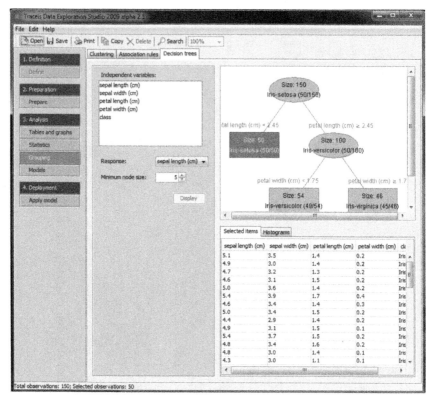

FIGURE B.27   Decision tree generated to classify the *Iris* plant species.

B.11.5 Exercise 4: Analysis of Census Data

This data set was collected from the 1994 census data and includes observations on individuals making either less than or greater than $50K per year. These measurements are captured in the variable *salary* from Kohavi & Becker (1994). There are 32,561 records with a set of variables containing measurements that include the following: *age*, *workclass*, *education*, *education-num*, and *occupation*.

The objective of this exercise is to characterize the differences between individuals making less than $50K and those making greater than $50K.

Step 1: Load the adult data set file *Adult.txt* located in the tutorials directory.

Step 2: Calculate new values for both the *education-num* and *age* variables that contain categories based on ranges, as shown in Figures B.28 and B.29.

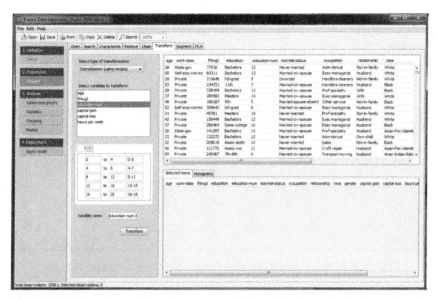

FIGURE B.28 Generation of a new variable with discretized values for the *education-num*.

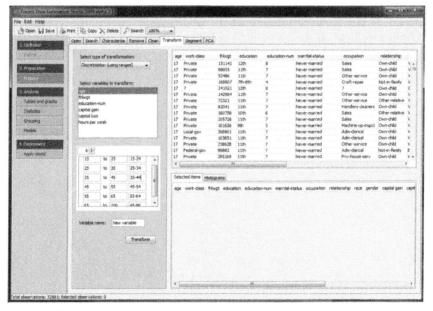

FIGURE B.29 Generation of a new variable with discretized values for the *age*.

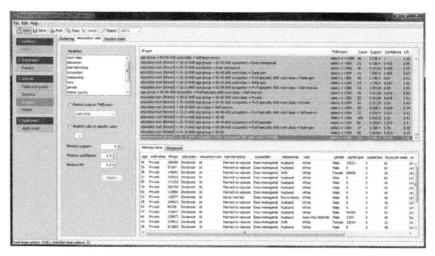

FIGURE B.30 Association rules generated from the adult data set.

Step 3: Generate association rules by selecting the "Association rules" tool, and selecting the following variables: *workclass*, *occupation*, *salary*, *age (discretized)*, and *education-num (discretized)*, with "Restrict rule on THEN-part" set to *salary*, "Minimum support" set to 15, "Minimum confidence" to 0.8, and "Minimum lift" to 1.0 (as shown in Figure B.30).

Step 4: Click on the individual rules to view the underlying data (as shown in Figure B.30).

BIBLIOGRAPHY

Agresti A (2013). *Categorical Data Analysis*, 3rd edn. John Wiley & Sons, Inc., Hoboken, NJ.

Alreck PL, Settle RB (2003). *The Survey Research Handbook*, 3rd edn. McGraw-Hill/Irwin.

Anderson DR, Sweeney DJ, Williams TA (2010). *Statistics for Business and Economics*. South-Western College Publishing.

Antony J (2003). *Design of Experiments for Engineers and Scientists*. Butterworth-Heinemann, Oxford.

Bache K, Lichman M (2013). *UCI Machine Learning Repository*, http://archive.ics.uci.edu/ml (accessed December 10, 2013). School of Information and Computer Science, University of California, Irvine, CA.

Baird J, Curry R, Reid T (2013). Development and application of a multiple linear regression model to consider the impact of weekly waste container capacity on the yield from kerbside recycling programmes in Scotland. *Waste Management & Research*, Vol. 31, pp. 306–314.

Barrentine LB (1999). *An Introduction to Design of Experiments: A Simplified Approach*. ASQ Quality Press, Milwaukee, WI.

Berkun S (2005) *The Art of Project Management*. O'Reily Media Inc., Sebastopol, CA.

Making Sense of Data I: A Practical Guide to Exploratory Data Analysis and Data Mining, Second Edition. Glenn J. Myatt and Wayne P. Johnson.
© 2014 John Wiley & Sons, Inc. Published 2014 by John Wiley & Sons, Inc.

Chapman P, Clinton J, Kerber R, Khabaza T, Reinartz T, Shearer C, Wirth R (2000). ftp://ftp.software.ibm.com/software/analytics/spss/support/Modeler/Documentation/14/UserManual/CRISP-DM.pdf (accessed December 10, 2013).

Cochran WG, Cox GM (1999). *Experimental Designs,* 2nd edn. John Wiley & Sons, Inc.

Cristianini N, Shawe-Taylor J (2000). *An Introduction to Support Vector Machines and Other Kernel-Based Learning Methods.* Cambridge University Press.

Dasu T, Johnson T (2003). *Exploratory Data Mining and Data Cleaning.* John Wiley & Sons, Inc., Hoboken, NJ.

Donnelly RA (2007). *Complete Idiot's Guide to Statistics*, 2nd edn. Alpha Books, New York.

Everitt BS, Landau S, Leese M, Stahl D (2011). *Cluster Analysis*, 5th edn. John Wiley & Sons, Ltd.

Fausett L (1993). *Fundamentals of Neural Networks: Architecture, Algorithms, and Applications.* Pearson.

Fielding AH. (2007). *Cluster and Classification Techniques for the Biosciences.* Cambridge University Press.

Fisher RA (1936). The use of multiple measurements in taxonomic problems. *Annals of Eugenics*, Vol. 7, No. 2, pp. 179–188.

Fowler FJ (2009). *Survey Research Methods (Applied Social Research Methods)*, 4th edn. Sage Publications, Inc., Thousand Oaks, CA.

Freedman D, Pisani R, Purves R (2007). *Statistics*, 4th edn. W.W. Norton, New York.

Gold target data (1999). http://www.stat.ufl.edu/~winner/data/gold_target1.dat (accessed December 10, 2013).

Guidici P, Figini S (2009). *Applied Data Mining for Business and Industry (Statistics in Practice).* John Wiley & Sons, Ltd.

Han J, Kamber M, Pei J (2012). *Data Mining: Concepts and Techniques*, 3rd edn. Morgan Kaufmann Publishers.

Hand DJ, Mannila H, Smyth P (2001). *Principles of Data Mining.* MIT Press.

Hastie T, Tibshirani R, Friedman JH (2009). *The Elements of Statistical Learning: Data Mining, Inference, and Prediction.* Springer, New York.

Hosmer DW, Lemeshow S, Sturdivant RX (2013). *Applied Logistic Regression*, 3rd edn. John Wiley & Sons.

Hsu J (1996). *Multiple Comparisons: Theory and Methods.* Chapman & Hall/CRC.

IRIS Flower data (1936). http://archive.ics.uci.edu/ml/datasets/Iris (accessed December 10, 2013).

Jackson JE (2003). *A User's Guide to Principal Components.* John Wiley & Sons, Inc., Hoboken, NJ.

Jolliffe IT (2002). *Principal Component Analysis*, 2nd edn. Springer-Verlag, New York.

Kachigan SK (1991). *Multivariate Statistical Analysis: A Conceptual Introduction*, 2nd edn. Radius Press, New York.

Kerzner H (2013). *Project Management: A Systems Approach to Planning, Scheduling and Controlling*, 9th edn. John Wiley & Sons.

Kimball R, Ross M (2013). *The Data Warehouse Toolkit: The Complete Guide to Dimensional Modeling*, 2nd edn. Wiley Publishing Inc., Indianapolis, IN.

Kleinbaum DG, Kupper LL, Nizam A, Rosenberg ES (2013). *Applied Regression Analysis and Other Multivariate Methods*, 5th edn. Cengage Learning.

Kohavi R, Becker B (1994). http://archive.ics.uci.edu/ml/datasets/Adult (accessed December 10, 2013).

Levine DM, Stephan DF (2010). *Even You Can Learn Statistics: A Guide for Everyone Who Has Ever Been Afraid of Statistics*, 2nd edn. Pearson Education, Inc.

Lindoff G, Berry MJA (2011). *Data Mining Techniques for Marketing, Sales and Customer Support*, 2nd edn. Wiley Publishing, Inc., Indianapolis, IN.

Montgomery DC (2012). *Design and Analysis of Experiments*, 8th edn. John Wiley & Sons, Inc.

Myatt GJ, Johnson WP (2009). *Making Sense of Data II: A Practical Guide to Data Visualization, Advanced Data Mining Methods, and Applications*. John Wiley & Sons.

Oppel A (2011). *Databases DeMYSTiFieD*, 2nd edn. McGraw-Hill/Osborne.

Pearson RK (2005). *Mining Imperfect Data: Dealing With Contamination and Incomplete Records*. Society of Industrial and Applier Mathematics.

Project Management Institute (2013). *A Guide to the Project Management Body of Knowledge (PMBOK Guides)*, 5th edn. Project Management Institute.

Pyle D (1999). *Data Preparation for Data Mining*. Morgan Kaufmann.

Rea LM (2005). *Designing and Conducting Survey Research: A Comprehensive Guide*. Jossey-Bass.

Rohanizadeh SS, Moghadam, MB (2009). A proposed data mining methodology and its application to industrial procedures. *Journal of Industrial Engineering*, Vol. 4, pp. 37–50.

Rud OP (2000). *Data Mining Cookbook: Modeling Data for Marketing, Risk, and Customer Relationship Management*. Wiley.

Sahoo NR, Pandalai HS (1999). Integration of sparse geologic information in gold targeting using logistic regression analysis in the Hutti-Maski Schist Belt, Raichur, Karnataka, India—a case study. *Natural Resources Research*, Vol. 8, No. 3, pp. 233–250.

Scottish recycle data (2013). http://www.stat.ufl.edu/~winner/data/scottish_recycle.dat (accessed December 10, 2013).

Tufte ER (1990). *Envisioning Information*. Graphics Press.

Tufte ER (1997a). *Visual Explanations: Images and Quantities, Evidence and Narrative*. Graphics Press.

Tufte ER (1997b). *Visual and Statistical Thinking: Displays of Evidence for Making Decisions*. Graphics Press.

Tufte ER (2001). *The Visual Display of Quantitative Information*, 2nd edn. Graphics Press.

Tufte ER (2006). *Beautiful Evidence*. Graphics Press.

Urdan TC (2010). *Statistics in Plain English*. Routledge, Taylor & Francis Group.

Vickers A (2010). *What Is a p-Value Anyway? 34 Stories to Help You Actually Understand Statistics*. Addison-Wesley.

Westfall P, Hochberg Y, Rom D, Wolfinger R, Tobias R (1999). *Multiple Comparisons and Multiple Tests Using the SAS System*. SAS Institute.

Winner L (2013). http://www.stat.ufl.edu/~winner/data/cruise_ship.dat (accessed December 10, 2013).

Witte RS, Witte JS (2009). *Statistics*, 9th edn. Wiley.

INDEX

Making Sense of Data I: A Practical Guide to Exploratory Data Analysis and Data Mining,
Second Edition. Glenn J. Myatt and Wayne P. Johnson.
© 2014 John Wiley & Sons, Inc. Published 2014 by John Wiley & Sons, Inc.

CPSIA information can be obtained at www.ICGtesting.com
Printed in the USA
BVOW06s0625030815

411020BV00006B/7/P